Everything Evolves

Everything Evolves

WHY EVOLUTION EXPLAINS
MORE THAN WE THINK,
FROM PROTEINS TO POLITICS

MARK VELLEND

PRINCETON UNIVERSITY PRESS
PRINCETON & OXFORD

Published by Princeton University Press
41 William Street, Princeton, New Jersey 08540
99 Banbury Road, Oxford OX2 6JX

press.princeton.edu

GPSR Authorized Representative: Easy Access System Europe - Mustamäe tee 50, 10621 Tallinn, Estonia, gpsr.requests@easproject.com

All Rights Reserved

ISBN 978-0-691-25340-4
ISBN (e-book) 978-0-691-25343-5

British Library Cataloging-in-Publication Data is available

Editorial: Alison Kalett and Hallie Schaeffer
Production Editorial: Jenny Wolkowicki
Jacket design: Drohan DiSanto
Production: Danielle Amatucci
Publicity: Matthew Taylor
Copyeditor: Anita O'Brien

This book has been composed in Arno Pro

Printed in the United States of America

10 9 8 7 6 5 4 3 2 1

CONTENTS

PREFACE

When the Covid-19 pandemic triggered lockdowns around the world in the spring of 2020, I was scared and anxious like everyone else. I was also feeling a bit guilty. With friends and loved ones healthy, children old enough to not need constant supervision, and a comfortable living space, there wasn't much to do but stay at home and make the best of it. And for someone who loves to read and to ponder, the prescription to hunker down at home, at least for a while, hardly felt like punishment. No daily commute, few administrative meetings, no lifts to school, no weekend drives to distant hockey arenas. With a sharp twinge of guilt, it was hard not to feel a bit liberated. So I read. A lot.

Despite the physical isolation, I didn't read alone. With friend and colleague Françoise Cardou, we formed the two-person Sherbrooke Critical Reading Society, with regular online meetings. One of our recurring topics was the widespread application of evolutionary ideas, and on March 18, 2020, we created a shared document for notes on various papers and books. Some of the early entries included the books *Diversity and Complexity* (Scott Page) and *The Evolution of Everything* (Matt Ridley), and a series of papers on cultural homogenization and globalization. Entry number 6 is my own previous book, *The Theory of Ecological Communities*, in which I had applied concepts from evolutionary biology to understand communities of many interacting species. There was a lot to read, but we were getting a sense that there were some things missing from the literature.

We learned that for well over 150 years, scholars have been noting a common set of processes at work in the evolution of all things cultural and biological. Technologies—whether sailboats or cell phones—did not appear in their current forms like flashes of lightning in an inventor's mind. Human languages and biological species were not created from scratch by a divine entity. All these things have ancestors near and distant, and a long history of trial and error by which new variants arose, had varying degrees of success, and passed down some of their characteristics to future generations. That is, they evolved. When someone says that Steve Jobs invented the iPhone, what they really mean is that he oversaw a team of people assembling dozens of existing and already sophisticated technologies into a slightly different combination. Actually, the team tried many combinations, winnowing down the possibilities with countless rounds of trial and error. It was one small episode—consequential as it was—in a long evolutionary trajectory.

One of the most provocative assertions about the generalized notion of evolution was made by McGill University biologist Graham Bell, from whom I learned evolution as an undergraduate. His book *Selection: The Mechanism of Evolution* starts with a one-page section titled *The Second Science*. After describing physics as the science most familiar to most people, he goes on to say:

> It fails to provide any understanding of how living organisms become adapted to their conditions of life; this is the province of the second science, the science of evolution. . . . No knowledge of physical principles, no matter how profound or detailed, can lead to any understanding of evolution, and vice versa. . . . So far as we know, the whole of the natural world can be understood in terms of just two general systems, and no more.

It bears repeating: The whole of the natural world can be understood in terms of just two general systems, *and no more*. Consider me provoked. Bell describes a fence that divides "different scientific modes of inquiry." On one side are the disciplines whose explanations all trace back to physical laws—including physics, chemistry, neurobiology, and physiology—for which he never uses the moniker the "First Science," although it is clearly implied. On the other side of the fence is everything from evolutionary biology and ecology to economics and history, "whose distinctive feature is the operation of selection on variable populations." This is the Second Science.

If it's true that physics and evolution are not just two big ideas in science, but *the only two things* you need in order to understand how everything came to be, why isn't this the first thing taught to students of science? The quick and easy answer is that the assertion is just not true. But if you think about evolution in its generalized sense, it becomes impossible (for me anyway) to think of a molecule, water body, planet, animal, language, culture, economy, or technology whose form and function cannot be explained by some combination of physics and evolution. So, let's say it's true. In that case, assuming that people already know about the First Science, surely they need to hear more about the Second Science.

There have, not surprisingly, been previous arguments in favor of a generalized evolutionary science, albeit without using the term *Second Science*. I have taken inspiration, and learned a tremendous amount, from related books by Marion Blute, Gary Cziko, Geoffrey Hodgson and Thorbjørn Knudsen, Eva Jablonka and Marion Lamb, Alex Mesoudi, Matt Ridley, Shiping Tang, David Sloan Wilson, and others. The book you are reading now is their descendant and was born from the hypothesis that a few factors have held back the Second Science from reaching the broad audience it deserves. I have attempted to overcome these limitations by writing a book with the following three attributes: First, it is accessible to anyone with some curiosity about the workings of nature and culture, not just to other academics. Second, it doesn't push any political messages. Third, it puts all subdisciplines within the Second Science on an equal footing, rather than placing biology in a position of primacy, from which we must look for direct analogies. You will be the judge of whether I succeeded.

There are many people whose assistance I need to acknowledge, and my gratitude goes first and foremost to Françoise Cardou. She helped develop some of the core ideas, provided a sounding board for a zillion tangential thoughts, called me out for overuse of clichés or jargony phrases or for being overly defensive, and provided a rare kind of brutally honest but always constructive feedback. I accept responsibility for all remaining transgressions, and I hope to have at least fixed the parts where she suggested I write "maybe something that doesn't sound like you're talking to ten-year-olds." In short, this book would not have been written without Françoise's input and support. Merci beaucoup.

Our reading club started during the Covid-19 pandemic, which was itself a vivid illustration of evolution happening in real time on many levels. Newspaper articles described the evolution of the virus itself, in terms of specific DNA mutations that altered the efficiency of transmission or the severity of symptoms. In the very same issues of the same newspapers, other articles described how governmental and societal responses, from school closures and reopenings to mask mandates and social distancing, were evolving by trial and error. These two examples of evolution—one biological, one cultural—give a glimpse of what this book is about, by illustrating the range of phenomena that are subject to evolutionary change: everything from organic molecules (like DNA and proteins) to social policy and politics. This means that I had to read well beyond my normal comfort zone as an ecologist, and that I had to consult colleagues and friends with expertise on a wide range of topics.

During the research phase for this book, many experts helped clarify technical details or aspects of how their disciplines work. These people include Diane Hagglund, Peter Moffett, Jean-Marc Patenaude, Fanie Pelletier, Sébastien Rodrigue, Tiziano Squartini, and Michael Stratigos. Invaluable feedback on earlier versions of the ideas or the text was provided by Madelaine Anderson, Graham Bell, Heather Berry, Anna Crofts, Eugene Earnshaw-Whyte, Julien Eychenne, Jeremy Fox, Hasanki Gamhewa, Kate Kirby, Stefan Linquist, Harriet Neal, Ming Ni, Jon Norberg, Sabine St-Jean, Guillaume Tougas, Chris Thomas, Annabelle Vellend, Christine Wallis, and two anonymous reviewers. Even short and seemingly minor comments sometimes turned out to be quite important (in case someone is surprised to see their name in the list). I owe thanks to the Université de Sherbrooke for permitting faculty to take sabbaticals, without which I could never have imagined writing an entire book. I saved who knows how many hours of time thanks to the people who built and maintain the Internet Archive (archive.org) and Project Gutenberg (www.gutenberg.org), precious resources for consulting historical texts online. The librarians at the Université de

Sherbrooke handled all my many interlibrary loan requests with a smile, and several fellow researchers shared publications I couldn't otherwise access: Lindell Bromham, Sylvain Gandon, Mahdi Kafaee, and Felicity Meakins. Sage advice on the process of writing for a general audience was provided by Anurag Agrawal, Paige Harden, and Alison Kalett. At every stage of the process, from vague idea to finished manuscript, Alison has been an expert guide and an incisive editor.

I need to give a special thanks to my daughter Annabelle. Early on, I was given the advice of imagining one specific reader who somehow represents the broader readership and then writing for them. But I struggled to identify that person, until I realized she was living under the same roof. I envision the reader of this book as someone who has learned enough to have some appreciation for the sciences—both social and natural—without necessarily being an expert. They are interested in how *everything evolves*: from genes, proteins, and species to languages, technologies, political systems, and culture. They might know a great deal about some of these topics, or almost nothing, but they are curious to know more. That's Annabelle, and hopefully it's you too. Annabelle's comments on an earlier version of the manuscript told me which bits would likely appeal only to other academics (those bits were chopped), which ones were laden with unnecessary jargon, and which explanations were overly convoluted (those bits were revised). She also drew the lovely illustrations you will see in several of the figures.

My hope is that this book will represent one incremental step on an evolutionary trajectory that sees the Second Science increasingly appreciated as one of the pillars on which our understanding of nature and culture stands. But more important, I hope you enjoy reading it.

Mark Vellend
August 26, 2024
Sherbrooke, Canada

1

Introduction

To my childhood ears, my dad sounded like an alien. While my brother and I were wrestling in the living room or eating breakfast, our father would pick up the phone and proceed to vocalize a string of mysterious sounds. From overhearing these conversations, we eventually figured out that *ema* meant mother (he was talking to Grandma), and when Grandma scolded us with *paha poiss!* the meaning was clear enough (bad boy!). But otherwise, Dad and Grandma could carry on in total secrecy. They were speaking Estonian.

The Cheerios box on the breakfast table presented its own set of mysteries. A groggy, disheveled, hungry child who has just rolled out of bed tends to stare blankly over their cereal bowl at whatever is right in front of them. I must have read that ingredient list thousands of times. On one side of the box, it said oats, corn starch, sugar, and salt, along with a handful of other items with peculiar names. (I still don't know what tocopherol means without looking it up.) On the other side of the box—in Canada—it said *avoine entière, amidon de maïs, sucre,* and *sel.* I congratulated myself on identifying the connection between *sucre* and sugar, and between *sel* and salt, and I drew the obvious conclusion that knowing English might help me make sense of roughly half the French language.

My childhood adventures with language got stranger still on Friday nights at my aunt and uncle's house. For Estonian and French, I could see the same basic alphabet I knew from English. But when one of my cousins opened the Torah for a short reading before Shabbat dinner, not

1

only did I fail to understand the spoken words, I couldn't even guess at how to pronounce the Hebrew symbols on the page. It didn't help that they were being read right to left. I had memorized a decent bit of the short blessing phonetically (*Baruch Atah Adonai . . .*), but the sounds held no more meaning to me than the end of the Beatles' song "Hey Jude": *Naaa Naa Naa, Nanana Naaa.*

I like to think that even as a child I could have ranked these four languages based on how similar they are to one another. Starting with English, French is most similar, followed by Estonian, and then Hebrew. A good teacher could have even nudged me to organize them into a genealogical tree, much like I might have done for my siblings and cousins, tracing our ancestry back to parents, grandparents, and so on. My language tree might have looked like figure 1.1. The visual immediately suggests the idea that once upon a time there were other languages from which the current set originated, and that the most recent ancestor shared by English and French (the rightmost circle) was spoken more recently than the common ancestor of all four languages (the circle at the left). The tree also suggests that languages must change drastically over time—enough to produce English and Hebrew from the same starting point—an idea that clashes with a child's sense that the vocabulary and grammatical rules of their mother tongue are fixed and enforced for all eternity by an evil cabal of teachers with red pens. So, what do the experts think?

Few topics have sparked the interest of researchers as much as language. As long as one is exposed to only one language, the *concept* of language goes largely unnoticed. But once exposed to multiple languages, it is near impossible not to feel a certain curiosity and wonder at the equally profound differences between languages (I can't understand a word those people are saying) and their essential sameness (they seem to be communicating just like I do). The differences can create frustration, but also a strong motivation to learn.

Sir William Jones was a prolific learner of languages. Born in London, England, in 1746, he went on to learn more than a dozen languages, including Greek, Latin, Hebrew, Sanskrit, and German, along with his native English and Welsh. Having studied languages as a child and as a

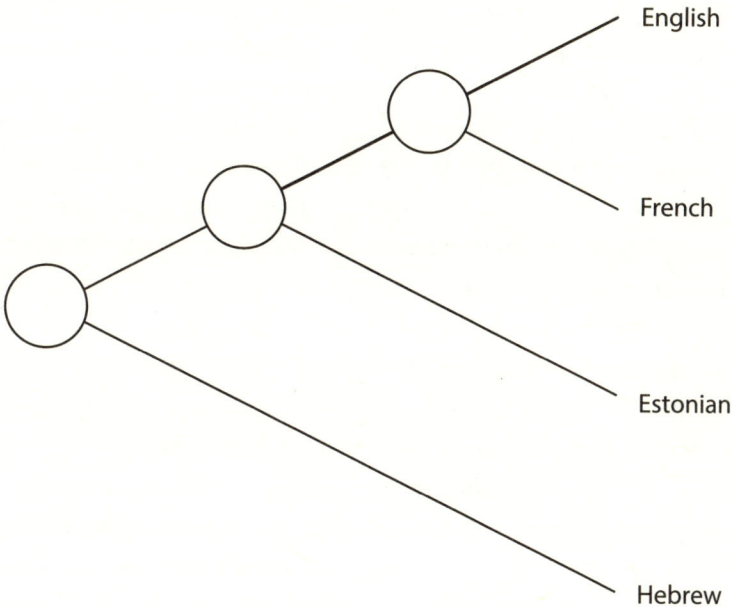

FIGURE 1.1. A child's-eye view of language similarities.

student at Oxford, Sir William later became a judge in Calcutta, India, and a reputed scholar of philology—the historical study of languages. His intimate knowledge and analysis of many languages led him to the striking insight that modern languages could indeed be traced back in time to common ancestors, a conclusion that ran counter to some contemporary ideas about languages being designed and fixed in time, divinely or otherwise. Speaking specifically of Sanskrit, Greek, and Latin, Jones said that "no philologer could examine them all three, without believing them to have sprung from some common source, which, perhaps, no longer exists."

Following Jones's lead, for the past two centuries linguists have analyzed how and why languages change over time. Some kinds of change follow systematic patterns—that is, they are more than random happenstance. For example, across hundreds of different languages, the most frequently used words tend to be the shortest (in English, think of *a*, *the*, and *of*), and the longest words are hardly ever used (try slipping "antidisestablishmentarianism" into your next conversation). This tendency is known as Zipf's law of abbreviation, named after the American linguist George Kingsley Zipf, who brought it to wide attention in the 1930s. Other kinds of consistent trends in language evolution had been identified in the first half of the nineteenth century by German philologists, such as Jacob Grimm (these are called Grimm's laws).

These regularities tell us that some language variants provide systematic advantages over others in terms of the efficiency of speech and communication. Other changes seem to be without rhyme or reason. When my Canadian ears hear a British person saying "lorry" or "trainers," it takes me a second to connect these words with "truck" and "running shoes." But there's no obvious advantage to wearing running shoes instead of trainers just because my feet are in North America. Still other changes can be traced to the movement of whole words or phrases from one language to another. When English was still the only language I could speak, my vocabulary nonetheless already included a few bits of French (*bon appétit, hors d'oeuvre*), a language I can now speak reasonably well, and even Yiddish (*klutz, schmooze*), which I never learned.

Individually, each of these observations about language is unremarkable. But if you put them together and look at them with blurred vision—abstracting from the details—in fact they contain all the ingredients needed to understand evolutionary change in a way that applies well beyond languages. When most people think of evolutionary change, they first think of Charles Darwin, who is credited with presenting the modern theory of evolution in his book *On the Origin of Species* (1859). As we shall see, however, the core ideas go back much further, and the fact that we draw a tight association between evolution and biology—instead of also including language and other aspects of culture—is due to a quirk of history. To see what I mean, let's rewind to the mid-nineteenth century for a thought experiment.

An Imaginary Scholar

Imagine a clever young mind circa 1850, digesting the observations about language just laid out. Intrigued by their sameness and differences, she is devising a theory of language change. How did we get from one or a few early languages to thousands of mutually incomprehensible tongues? What kinds of processes got us from the English in a Shakespeare sonnet to the English in a Charles Dickens novel? She was aiming for something that applied to any and all languages, and her theory had several components.

First, for meaningful change to happen in language, one or more sources of novelty are required. We need new variants for things to change. No problem: People experiment with language all the time. We are constantly trying out new pronunciations or meanings of old words, inventing words for new discoveries or technologies, altering the ordering of words in phrases, and so on. Before there were knives, engines, and bicycles, those words—or at least certain meanings of them—didn't exist. In short, there are plenty of ways in which people produce new language variants.

If language changes are to accumulate over time, rather than just disappearing as quickly as they appeared, we also need a means by which new variants can be passed on from one person to another. That is, we need a mechanism of inheritance. Learning fits the bill here. Everyone learns their mother tongue from parents, relatives, teachers, friends, acquaintances, and written materials (to which we can now add recordings and broadcasts). We can inherit language variants from any one of these influences, and then pass them on to our own children, friends, students, and so on. Each person has a slightly different combination of influences, so the version of a language that any one of us speaks will be at least slightly different from everyone else's.

Next, we need to account for why some variants in a language never catch on (the French term *chien chaud* didn't manage to displace the anglicized *hot dog* in Québec), why some variants quickly become widespread ("app" has now largely replaced "software"), and why some formerly common variants are no longer used outside of literature class ("wherefore art thou Romeo"). As we saw from the tendency for the most common words to become the shortest ones, some changes are driven by a process of systematic selection among different variants, given that some variants do a better job than others at facilitating communication (it's easier to say "can't" than "cannot"). Other changes appear entirely random with respect to serving any function, allowing languages to drift apart haphazardly. *Hello* (English), *tere* (Estonian), *bonjour* (French), and *shalom* (Hebrew) are effectively equivalent greetings, none better than the others, just different.

Finally, to generate a diversity of languages, we need for people to be clustered into sufficiently isolated groups to allow differences to accumulate between them. Geography does the job here. Geographically isolated groups of people will come up with different modifications to a language, and as those changes accumulate over time, multiple languages will emerge where previously there was just one. It is unlikely that English, German, and Dutch would have diverged into distinct languages without some geographic separation between the groups of people speaking early versions of each. In addition, a small amount of movement might permit languages to both remain distinct while also influencing one another, via the borrowing of words, phrases, or structural features.

So far, our hypothetical nineteenth-century prodigy has a theory of language change requiring just a few ingredients: constant sources of new variants, a means of inheritance, two types of internal dynamics (selection and drift), and movement between populations that is constrained to some degree. Taking the thought experiment a step further, we can imagine that she also started to note some features of her nascent theory that indicated she might be onto something applicable well beyond language. Could other aspects of culture or life change according to the same recipe? The ingredients show some striking similarities with factors thought to be important in other areas of inquiry that were well-known or just emerging at the time.

In the eighteenth century, Adam Smith had described efficient economic markets resulting from variation among people in what they needed and what they had to offer, and systematic adjustments in their behavior (selection among alternatives) that maximized individual net benefit. At the dawn of the nineteenth century, naturalists like Jean-Baptiste Lamarck in France were steering explanations of biological evolution away from the divine and toward the natural. Lamarck invoked new traits arising (e.g., a giraffe stretching its neck longer) that offspring could inherit, slowly allowing species to adapt to their environments and to diverge in form and function from other species. Plant and animal breeders were enjoying great success in identifying potentially useful varieties and then serially mating and selecting the "best" ones to produce potatoes or sheep that were higher-yielding or better-tasting. Reading Smith, Lamarck, and others would have revealed some striking parallels between the forces of change in language evolution and those at work in economic markets and in biological species.

By the time she caught wind of two British naturalists—Charles Darwin and Alfred Russel Wallace—in a race to get the word out on their ideas about biological evolution in the 1850s, our intrepid scholar had already devised a Generalized Theory of Evolutionary Change. The theory applies to any *evolutionary system*, which can be defined by three key ingredients: (1) entities that vary one to the next in some characteristics; (2) some means by which characteristics can be passed on over time; (3) a variable degree of success among entities in passing on their characteristics. Languages, technologies, markets, plants, and animals all fit the bill. Restricted movement and the consequent (partial) isolation of different parts of the system—based on geography or other barriers—are often involved, but they are not *necessary* ingredients. Evolution can happen even in one small place.

Since she was excluded from intellectual discourse at the time, our imaginary scholar's thoughts and notes were all kept to herself; they did not survive to the present day. As a result, the elegant idea of Darwin and Wallace—having emerged as the winner in the competition of ideas—came to be known as *the* theory of evolution. But over the years, many other scholars have seen what she saw: Darwin and Wallace's ideas are better described as constituting *a* theory of evolution—one specific application of the generalized version. Indeed, Darwin himself noted the parallels between biological change and language change, taking inspiration from linguistics for his own theory. The economist J. Stanley Metcalfe captured the situation nicely: "Evolutionary theory is a manner of reasoning in its own right quite independently of the use made of it by biologists."

The Darwinian Distraction

Although the elements of contemporary evolutionary theory go back many centuries, these elements had not, in fact, been assembled into a Generalized Theory of Evolutionary Change by the early nineteenth century. This is why our scholar is imaginary. Credit for the first coherent, compelling, and ultimately successful theory of evolution is almost universally attributed to Charles Darwin (usually given more credit than Wallace), whose biological theory of evolution by natural selection was published in its full form in 1859. Few ideas have had as profound an impact on science and society as Darwin's, not only by advancing our basic understanding of life on Earth, but also by permanently unseating humans from their presumed position of supremacy among living beings. However, it is quite easy to imagine the same essential theory having been first devised for language or technology, rather than biology. In chapter 2 we will delve into the history of evolutionary ideas across many disciplines. For now, a quick summary of the modern theory of biological evolution helps to illustrate what I call the Darwinian Distraction.

The biologists to whom Darwin passed his theoretical torch have built an elegant mathematical theory that has stood the test of time in impressive fashion. The modern version of evolutionary theory in biology, often referred to as neo-Darwinism, involves random mutations in DNA, a molecule unknown to Darwin. Stretches of DNA that provide instructions on how to make proteins are called *genes*, and proteins ultimately do most of the things in cells that give organisms characteristics like their size, shape, color, or metabolism (their "phenotype"). Genes are passed on to offspring, so any change in phenotype that benefits survival and reproduction (fitness), such as antibiotic resistance in the face of antibiotics, will spread in a population. In essence, nature selects the variants best suited to local conditions, so we call it *natural* selection. Because this theory is so well worked out and widely applicable, it has frequently created a strong force channeling thinking about other manifestations of evolution toward analogies with the biological version. Such analogies can sometimes be fruitful, but sometimes they can do more harm than good.

The temptation to make direct analogies from biological concepts such as the gene, genome, or phenotype to components of other evolutionary systems is hard to resist. Language is a good example. The building blocks of DNA are frequently referred to as letters in a four-letter alphabet: A, G, C, and T, for Adenine, Cytosine, Guanine, and Thymine. These are four types of nitrogen-rich base that distinguish each link in the DNA chain. So, it seems like a stretch of DNA is kind of like one long text, in which case it should be a short step from neo-Darwinism to understanding evolutionary change in strings of letters from a twenty-six-letter alphabet (along with punctuation and spaces). But then, what's the equivalent of a gene? Maybe a word . . . or a phrase . . . or a sentence? And maybe the phenotype is the *meaning* of the word or phrase or sentence? Maybe. Maybe not. You can probably sense that we're already holding on by a narrow thread. Try taking the analogies much further and the thread snaps.

To be sure, linguists have meticulously dissected languages into many component parts, but attempting to map phonemes, morphemes, or lexemes onto codons, genes, or chromosomes is unlikely to lead anywhere but to some laughs at the bar (if you're a linguist, anyway). And if the analogy project falls apart for language, it barely gets off the ground for information technology or religion, both of which undergo the same essential kind of evolutionary dynamics. In short, starting with neo-Darwinian concepts in biology and working out from there via analogy distracts us from seeing the true breadth of application of evolutionary ideas. It is no more important to find analogies for genes outside of biology than it is to find an analogy for verb tenses in a genome. Analogy projects are also misguided in assuming that key concepts have clean, unambiguous definitions in biology itself. As we shall see throughout this book, biology is itself quite messy.

Although attempts at making precise analogies often lead to dead ends, there must be *some* comparable features of languages, technologies, plants, and animals if a generalized theory is to apply to all of them. The trick is to find the level at which comparisons work, which is unlikely if we start with one particular kind of system (e.g., a neo-Darwinian biological one) and then insist on finding analogies elsewhere (e.g., looking for genes in technology). Within evolutionary biology, general theories do not make reference to specific characteristics of animals like their blood or their teeth, since there are no such things in plants, fungi, or bacteria, and we expect the theory to apply to all of them. Instead, blood and teeth are considered as two examples of characters or phenotypes, and it is these more abstract concepts that allow biological evolutionary theory to apply to any measurable attribute of any organism. For a generalized evolutionary theory (within which biology is but one example), some clues as to the appropriate level of abstraction and comparison can already be seen in the definition of an evolutionary system given earlier: We will at least need the concept of an "entity" and some measure of their "success." While languages and species both share the essential features needed for evolution, we don't need gene-like things in language, we don't need verb-like things in biology, and we don't need to begin with Darwin.

Beyond analogies, there is a second important element to the Darwinian Distraction. In the social sciences—where scholars study human social relationships and culture—there is great potential for generalized evolutionary thinking, but just the word *Darwinian* can create a major distraction. Following the publication of Darwin's theory, it was quickly seized on to support racist ideologies, with arguments along the lines of it being only "natural" for powerful white men to subordinate everyone else. The field of eugenics was born with the aim of improving the human race by placing restrictions on who was allowed to reproduce. Nazi Germany was the pinnacle of evil invocations of "social Darwinism." A superficial analysis would suggest that we must heed the social dangers inherent in evolutionary thinking. However, social Darwinist projects actually involve misrepresentations of evolutionary theory, invoked to advance particular political agendas. One can be adamantly opposed to racist ideology and in favor of generalized evolutionary theory.

For more than 150 years, scholars have been hinting at, noting, or formally analyzing the potential for generalized evolutionary theory. But almost all such efforts begin with a focus on Darwin. Evolutionary biologist Richard Dawkins proposed "Universal Darwinism," while others prefer "Generalized Darwinism," with book titles referring to "Darwin's conjecture," "Darwin's dangerous idea," "the second Darwinian revolution," or "how Darwinian theory can explain human culture and synthesize the social sciences." Subsequent discussions can become mired in distracting analogies or knee-jerk objections to any sentence combining "Darwin" and "social." This book is an attempt to break free from the Darwinian Distraction and to show that a generalized theory can be applied to all evolutionary systems. In the realms of life and culture, *everything* evolves.

Two Kinds of Science

Generalized evolutionary theory cuts across what has traditionally been the biggest and seemingly most fundamental disciplinary boundary in science: the one between the natural and the social sciences. The natural sciences include physics, chemistry, biology, and their extensions, such as astronomy, biochemistry, and ecology. The social sciences are about people, with subdisciplines that include sociology, economics, history, political science, and anthropology. Generalized evolutionary theory not only cuts across the natural-social divide, but it suggests a very different fundamental feature that distinguishes different branches of science.

Within the natural sciences, there is conceptual unity between physics and chemistry, because chemistry is essentially physics applied to molecules. This unity also extends part way into biology: If you want to know what makes your muscles feel sore the morning after a day of intense exercise (physiology), how lizard feet stick to walls (biophysics), or how plants make sugar out of air and water (molecular biology), the answers will be expressed in terms of physics and chemistry. Other parts of biology fall outside of this sphere of unity. If you want to understand how the world came to have humans, lizards, or plants in the first place, physics and chemistry are of little help. What you need for that is evolution: the process by which the generation of variation, inheritance, and differential success lead to exquisite adaptations. And as it turns out, the very same general process of evolution—this time in the social and cultural realm—explains how human societies came to have the English language (linguistics), gender roles (sociology), central banks (economics), democracy (political science), and iPhones (technology). In other words, there is a more meaningful divide in science than the natural-social one. On one side, there are questions that can be answered entirely with reference to physical processes. On the other side are questions for which answers require the addition of evolutionary processes.

So, if evolution is the key distinguishing feature of systems that can't be reduced to physics alone, then perhaps there are two—and only two—truly fundamental branches of science. Physics can be considered the first branch, held up by historians and philosophers as the paragon of scientific achievement. Physicists and chemists have identified a set of laws from which they can understand and predict the behavior of everything from planets and continents to atoms and subatomic particles. We can call this the "First Science." The second fundamental branch of science is the science of evolution, applying to everything from coronaviruses to computers. Biologist Graham Bell has dubbed this the "Second Science." The proposition is that physics and evolution together can explain everything. But in explaining how life and its products came to be as they are now, the First Science has but a supporting role to play. The lead actor in that play is the Second Science.

Even if most people (this author included) do not fully understand the laws of physics, especially at a subatomic level, no one can say that the universality of these laws is underappreciated. With Albert Einstein as the personification of science and genius, the First Science gets plenty of limelight. The Second Science, not so much. To be sure, Charles Darwin occupies rarefied air like few others in the scientific pantheon, but the idea of evolutionary systems is mostly seen to be left behind once you exit the confines of biology. As a biologist, my own students have learned evolutionary theory as it applies to biology, but they are mostly unaware that folks on the other side of our university campus apply the generalized version of evolutionary theory to culture, language, and economics. Despite many decades of formal and informal application of evolutionary ideas, well beyond the borders of biology, the Second

Science—as a unified branch of science—remains in the shadows. The Second Science deserves greater recognition as one of the two pillars holding up the entire scientific enterprise.

If one is accustomed to thinking about scientific theories and models from the point of view of the First Science, the Second Science might seem a bit odd. Mathematical models in physics produce some long-term predictions of astonishing accuracy. In 1705 Edmond Halley used theories of gravity and planetary movement to correctly predict that one particular comet orbiting the sun—now known as Halley's Comet—would return to viewing distance from Earth in 1758. Using the same theories, we now anticipate the next closest visit of Halley's Comet to our sun on July 28, 2061. Evolutionary theory, in contrast, does not have a lot to say about the specifics of what will be happening on that day. We could say that between now and then, some of the currently rarest species will go extinct, while other species will look pretty much like they do now, notwithstanding some minor evolutionary changes. New technologies and technology companies will have arisen, displacing old ones, with products that are faster and more powerful. Teenagers will communicate using platforms that don't currently exist. While evolutionary models can make some very useful short-term predictions (e.g., the rate of spread of a new virus), over the long term, predictions get increasingly uncertain.

Evolutionary dynamics are fundamentally unpredictable because each step in the evolutionary process is contingent on the previous steps, and there are an astronomical number of possible pathways that can be taken. Consider the virus that causes Covid-19 (called SARS-Cov-2), which is quite simple as evolutionary systems go, with a genome of about 30,000 DNA letters. With four letters in the DNA alphabet, each one can be changed from its current state in three ways. So, if we randomly choose just one letter and switch it to one of the other three, there are 90,000 possible DNA sequences for the new genome. The next DNA change also has 90,000 possibilities, so looking two mutations ahead, there are 8.1 billion possibilities ($90,000 \times 90,000$). Similarly, with roughly 500,000 words or phrases in the English language (according to the *Oxford English Dictionary*), a sequence of two random changes to word or phrase meanings also generates billions of trajectories ($500,000 \times 500,000 = 250$ billion). Evolution is fundamentally a process of trial and error, and it's hard to know what will be tried and which trials will stick around (or not).

So, if evolutionary theory doesn't make precise predictions about the distant future, what does it do exactly? Two things. First, the Second Science has excelled at teaching humanity the critically important lesson that the dynamics of biological and social systems involve some fundamental unpredictability in the long term, frustrating as this lesson might be to a physicist. Second, and more important for this book, the Second Science provides a unified theory for understanding how and why evolutionary systems change in general, for explaining what has happened in the past, and for anticipating, to some degree, possible changes in the future (with greater uncertainty the further you look ahead). Evolutionary systems include everything from languages and computers to viruses and whales, so achieving theoretical unification has not been easy. As mentioned earlier, the key has been to find the right level of abstraction, which requires a focus not on the differences among such diverse systems but on their commonalities.

The Second Science explains how we came to have not only eyes but also iPhones. It explains how different kinds of animals came to have different kinds of eyes, and why cell phones show so much diversity in form and function. In short, the Second Science helps us make sense of two of the most striking features of life and its products: their remarkable degree of adaptation for some function, and their diversity.

Everything Evolves: The Second Science Is Everywhere

The importance of the Second Science goes far beyond its role in underpinning our fundamental understanding of life and culture. In ways that are sometimes obvious, sometimes obscure, evolution lies at the core of countless human endeavors of profound importance. Whether consciously or not, people have been applying evolutionary principles for millennia. We have transformed wild plants and animals into the domesticated forms that make up most of our food supply. Cornfields in the United States now yield 700% more bushels per acre than they did one hundred years ago, a staggering increase achieved by selective hybridization and breeding (biological evolution under human selection) and the cultural evolution of agricultural practices and technology. We have produced some microbes that are little factories for the production of antibiotics or for the neutralization of harmful pollutants, and others that can eliminate pathogens that have evolved resistance to antibiotics.

Evolutionary principles are applied when developing all new technologies, whether a better kind of hammer or a space shuttle. The widespread idea that such technologies spring fully formed from the minds of lone geniuses is badly misguided. All such technologies have involved countless rounds of trial and error and gradual refinement via a profoundly evolutionary process: Ideas are generated, tested, and passed down to the next generation of inventors. The advancement of science is itself an evolutionary process, with each new theory built on the successes and failures of a long line of predecessors. Charles Darwin's theory of biological evolution by natural selection is a prime example, as is the perspective put forth in this very book. As the chemist Leslie Orgel once famously said, "Evolution is cleverer than you are."

One of the defining challenges for humanity in the twenty-first century will be grappling with the power of artificial intelligence, which is at its core an evolutionary algorithm. Artificial intelligence starts with scientists creating evolutionary systems in silico, allowing computers to evolve solutions to the thorniest of challenges, giving us self-driving cars, facial recognition software, and chatbots. Their level of "intelligence" can only be described as spooky.

With the accelerating movement of people, ideas, and species across the globe, the diversity of life and culture found in ecosystems, in cities, and in countries is changing rapidly. The Second Science guides us to an understanding of not only how and why diversity changes over time, but also what the consequences of such diversity might be for the productivity of economies or the stability of ecosystems. Evolutionary trajectories in economies, ecosystems, and political systems can sometimes involve tipping points—massive and difficult-to-reverse changes in response to a modest stimulus. Sparking and promoting social and political change and identifying tipping points we might like to traverse or avoid are also, at their core, applications of evolutionary principles.

All these topics—and many more—will be explored throughout this book. Their unifying feature is the same underlying process of evolution. The scientists, engineers, agronomists, politicians, businesspeople, health practitioners, and computer programmers working on these topics have a great deal to gain from recognition of a unified Second Science. Lessons learned in one corner of the Second Science (e.g., artificial intelligence) might have important lessons for others (e.g., selective breeding), but without a common language and conceptual framework, those lessons can go unlearned. In some parts of the world, there has been strong resistance to the teaching of evolution in biology, given a conflict between the idea of gradual evolution and the biblical account of the creation of life. In keeping evolution out of the classroom, we are robbing children not only of an understanding of how life works but also of one of the two fundamental kinds of process in science—the one that produces change in biological species, economies, cultures, technologies, and societies.

The rest of the book is structured as follows. Chapter 2 will establish the historical depth and disciplinary breadth of evolutionary thinking, from ancient history to the present, in disciplines ranging from linguistics and the philosophy of science to economics and biology. In chapter 3 I will more explicitly define evolutionary systems and establish the key concepts underlying the Second Science and their breadth of application. Next, we will delve into the core processes underlying the dynamics of all evolutionary systems: origination of variation, inheritance, selection, randomness, and movement (chapters 4–6). These chapters will lay out the essentials of the Second Science. The next three chapters will explore the profound importance of evolutionary thinking and understanding for human concerns and endeavors. Specific applications include artificial intelligence, animal and plant breeding, identification of system tipping points, and the causes and consequences of diversity of all kinds. Chapter 10 will step back and apply a wide-angle lens to seeing the importance of the Second Science and how it fits into the scientific enterprise as a whole.

Big-picture appraisals of science invariably gravitate toward physics as the epitome of success, with its universally applicable laws and models. Outside of physics is a jungle of more narrowly applicable and partially overlapping theories, among which we struggle to see common threads. Generalized evolutionary theory suggests that there is indeed a common core to all this. To understand the game of life and all its products, the First Science establishes the constraints, but the Second Science sets the rules.

PART I

Setting the Stage for the Second Science

2

The Evolution
of Evolutionary Thinking

Science begins with unsolved puzzles. One of my favorites is the question of where the material comes from to build a 10-ton tree from a 10-milligram seed—a true mystery in the eighteenth century. If we start with several hypotheses about how to solve this puzzle, the next step is to confront each of them with evidence that has the potential to prove them wrong. By weighing the soil before and after a plant is grown in a pot, we can reject the hypothesis that its mass is derived from the soil. After also rejecting water as the key source of a plant's mass, only one hypothesis is left standing: Plants are mostly made of air. But this was difficult to prove, in addition to running counter to most people's intuition.

According to the twentieth-century philosopher Karl Popper, the experiments just described are just one example of how science is done in general: We are not in the business of *proving* anything, perplexing as that might seem to a student starting out in science (as it was for me). Instead, the job of a scientist is to attempt to *disprove* or *falsify* proposed explanations or hypotheses about how the world works.

For a long time, I was under the impression that Popper's definition of science was *the* definition that "everyone" agreed on. But even in our example of tree growth, there is a problem. Given current technology, we can now effectively "see" carbon dioxide moving from the air into a leaf, where energy from the sun uses it to fuel the construction of sugars, which go on to build roots and wood and leaves. So, it is reasonable to say that we now *know*, or that scientists have *proven*, the astonishing fact that plants are built mostly from material grabbed out of the air, just as we now know that the Earth is shaped more like a ball than a pancake. In the end, Popper's view is insufficient to capture the full complexity of science. But it does capture the essential truth that scientific advancement involves repeated rounds of trial and error—that is, science is an *evolutionary* process. New ideas are continually put forth, and then empirical evidence determines the ultimate success or failure of those ideas.

Although the evolutionary view of science is often described as Popperian, the basic idea goes back much further. William Whewell was a highly influential English philosopher, scientist, and historian in the first half of the nineteenth century, with a complex relationship to evolutionary thinking. He described science as a process of trial and error in which a great many hypotheses are put forth, most of which are discarded in favor of the one that best aligns with the facts. Whewell's description of science showed a clear appreciation of the three key ingredients of an evolutionary process. Two ingredients were described explicitly: variation among hypotheses in their characteristics (he refers to "many possibilities," "many candidates"), and their differential success (some get "rejected," "discarded," or "condemned"). The third ingredient, inheritance, is implicit: Each "truth" joins our collective scientific knowledge and forms the basis for subsequent rounds of hypothesis formulation and testing.

These brief excerpts from the philosophy and history of science help illustrate some of the key themes of this chapter. In Whewell's description of science, we see that the core idea of evolution, involving the differential success and inheritance of variable entities, has arisen outside of biology and predates their application to biology. We can also see that science itself might provide a particularly vivid case study of evolution in action. This chapter can thus be read both as a concise history of evolutionary ideas and as an illustration of the inner workings of evolution in the world of ideas and culture.

While I argued in chapter 1 that an obsession with Charles Darwin and neo-Darwinism is actually an impediment to the advancement of generalized evolutionary science, the *historical* importance of Darwin's influence is unquestionable. On one hand, as described in the book *On the Origin of Evolution*, Darwin's "contribution was one link in a chain that extends back into antiquity, and is still being forged today." On the other hand, Darwin's skill in synthesizing and communicating ideas to audiences far beyond the dusty halls of universities and scientific societies meant that the publication of *On the Origin of Species* in 1859 sparked a rapid diffusion of evolutionary ideas that had a profound influence on science and society. With this in mind, the story presented here is organized by describing contributions before versus after 1859, and within biology (as we now understand that field) versus outside of biology.

The Deep Roots of Evolutionary Ideas in Biology

If we wrap generalized evolutionary thinking into a package and call it a single hypothesis, we can ask: What have been the main alternative hypotheses historically? In biology, the main alternative has been providence: God created each species separately, explaining both the adaptive fit of organisms to their circumstances (e.g., big beaks to eat big seeds) and their spectacular diversity (e.g., many thousands of bird species). Doubts about this providential explanation have been around for a long time. The origins of contemporary evolutionary thinking—including the idea that species change over time, and the idea of natural selection—have been traced back to the philosophers of antiquity, including Empedocles (fifth century BC), Zhuangzi (fourth century BC), Epicurus (fourth and third centuries BC), and Lucretius (first century BC), and to Islamic writers such as al-Jahiz (AD 776–868). Modern evolutionary ideas have unambiguous precedents dating back more than two thousand years.

Evolutionary ideas changed relatively little during the two millennia between antiquity and the Age of Enlightenment in the seventeenth and eighteenth centuries. Picking up the story in this era, philosophers and naturalists were grappling with a massive accumulation of facts about life, past and present, most notably fossil remains of unknown creatures in puzzling places. How did the rocks on mountaintops come to contain fossils of marine molluscs known as ammonites, with no living representatives? Why would God choose to create species like woolly mammoths, only to then let them go extinct? If species were created as unchangeable entities, why was it so easy for people to breed dogs with no hair or pigeons with feathered feet?

Before thinking too hard about *how* evolution worked, one needed to establish that evolution happens in the first place. The prevailing idea that species characteristics were fixed for all time—that they were immutable—had a strong hold on the collective imagination, but it was challenged by many philosophers and naturalists, such as Erasmus Darwin (1731–1802, Charles's grandfather) and Jean-Baptiste Lamarck (1744–1829). The alternative to species immutability came to be described as "transformism" or "transmutation," according to which the characteristics of a species could transform or transmutate over time, possibly resulting in new species.

The idea that one species can transform into another does not necessarily imply that all species trace their ancestry back to a common origin. Perhaps new species are being continually created spontaneously, albeit with the capacity for adaptation to their local environments via transformation (what we would now call evolution). So, the idea of common ancestry creating a single tree of life required its own development. For Christianity, an especially problematic implication of a single tree of life was the idea that humans evolved from "lower" forms of life—sharing a common ancestor not only with chimpanzees and gorillas, but ultimately also with fish, insects, and bacteria. Contradicting the biblical account of Creation was not a decision to be taken lightly. In the seventeenth century, Galileo Galilei spent the final nine years of his life (1633–1642) under house arrest for claiming that the Earth moved around the sun, rather than vice versa. Fear of the consequences of contradicting Holy Scripture led many evolutionary thinkers to delay publication, to write anonymously, to recant their published views, or to otherwise obscure their increasing certainty that the human species was just one branch among many on the tree of life.

As evidence pointed more and more forcefully to the mutability of species, the question of *how* evolution works gained in significance. The intricate design of a vertebrate eye or an orchid flower seemed to suggest that there must be a designer of some sort. The English priest William Paley (1743–1805) presented a compelling case for intelligent design, using the famous analogy of a watch, whose complex and fine-tuned design would appear to provide incontrovertible evidence of a designer. The absurdity of a watch assembling itself was then extended to apply to a frog being built without an intelligent Creator. Budding evolutionists faced a monumental challenge: Even if we accept that something like a fish can evolve into something like a human—without the assistance of a Creator—what *mechanism* can explain how and why this happens?

Jean-Baptiste Lamarck had an idea. He had observed that during their lifetimes, animals can change in size and shape based on the nature of their activities. If you use your bicep muscles more, they will get stronger. Probably the most famous (but ultimately fallacious) example is the giraffe stretching its neck to reach the treetops. Lamarck proposed that such alterations—acquired during the lifetime of an animal—could be inherited by its offspring, thus explaining how species change over time. If you lift weights, your children will have big muscles. Taken together, these ideas involve the three key ingredients of an evolutionary process: Some organisms are better than others at modifying their characteristics in response to environmental demands, and these characteristics are inherited by their offspring. *Selection* produces *variation*, which is then *inherited*.

Lamarck's theory of evolution was published in 1809. Twenty-two years later, in 1831, little notice was given to a new book with a riveting title: *On Naval Timber and Arboriculture: With Critical Notes on Authors Who Have Recently Treated the Subject of Planting*. Written by the farmer and forester Patrick Matthew, the book's appendices included some remarkable ideas on evolution, including an impressively clear description of what we now call natural selection:

There is a law universal in nature, tending to render every reproductive being the best possibly suited to its condition. . . . This law sustains the lion in his strength, the hare in her swiftness, and the fox in his wiles. As Nature, in all her modifications of life, has a power of increase far beyond what is needed to supply the place of what falls by Time's decay, those individuals who possess not the requisite strength, swiftness, hardihood, or cunning, fall prematurely without reproducing—either a prey to their natural devourers, or sinking under disease, generally induced by want of nourishment, their place being occupied by the more perfect of their own kind, who are pressing on the means of subsistence.

The very same year that Matthew's book came out, HMS *Beagle* sailed out of Plymouth, England, with a twenty-two-year-old naturalist named Charles Darwin onboard. Darwin did not bring a copy of (or know about) *On Naval Timber and Arboriculture*, but his extensive observations of plants and animals on the journey sowed the seeds of essentially the same idea. Darwin discovered the fossilized bones of extinct giant mammals and noted subtle but distinct differences between island and continental forms of many plants and animals. Most famously, the flora and fauna of the Galapagos Islands presented striking geographic patterns of variation, blurring the line between what we might call distinct species, or just varieties within species. From one island to the next, animals and plants were at once different . . . but also the same. When the *Beagle* set sail in 1831, Darwin subscribed to the conventional wisdom that species were immutable, rejecting the notion of transmutation. When the ship docked back in England in October 1836, he was not so sure.

Few episodes in the history of science have generated as much ink on paper as the inner thoughts of Charles Darwin between 1836 and 1859. Two points are key to the story I want to tell here. First, at some point in this period Darwin pulled together the various threads of evolutionary thinking into a coherent theoretical whole, culminating in a concise and forceful book: *On the Origin of Species*. Contrary to Lamarck, Darwin saw the generation of variation among organisms as a *random* process, with natural selection as the central creative force in evolution. In short, variation arises by random accident, some variants are better adapted to local circumstances and so leave more offspring than others, and those offspring inherit the characteristics of their parents. Innumerable repetitions of this cycle over millions or billions of years can gradually turn a fish into a human, without any need for an intelligent designer. If people take control of the middle step via artificial selection, they can turn a wolf into a poodle.

The second notable feature of the period between the *Beagle* and *Origin* is just how much time passed—more than twenty years—between when Darwin developed his core ideas and when they were finally published. Some argue that the delay can be explained, at least in part, by the heretical nature of the ideas: The theory showed remarkable alignment with the facts of nature, but equally remarkable misalignment with the teachings of the all-powerful Christian faith. Others see only evidence of an extremely busy and meticulous man taking the time needed to get the book done, amid many other projects. Either way, given Darwin's frequent health problems, he could easily have died prior to finally getting *Origin* published, showing that the evolution of science, like the evolution of everything else, involves lots of happenstance.

It is highly likely that Darwin would have further delayed the writing and publication of *Origin* if it weren't for a pesky young naturalist named Alfred Russel Wallace, who began his first overseas voyage in 1848 on a boat aptly named the *Mischief*. Unlike Darwin, Wallace began his travels already convinced of transmutation. But like Darwin, his observations of tropical plants and animals, especially in the Malay Archipelago, led him to the idea of natural selection as the key mechanism of evolution. In 1858 Wallace sent Darwin an essay in which Darwin immediately recognized the same idea he had been developing for more than twenty years. Later the same year, Wallace's essay was presented alongside one of Darwin's to the Linnean Society of London, as a first public announcement of the theory on which they independently converged.

When I first read *On the Origin of Species*, my university course in evolution had led me to expect long passages about the finches of the Galapagos Islands. I found no such passages. Instead, the reader of *Origin* is immediately bombarded with facts about domestic pigeons. The book includes intricate descriptions of pigeons' feet, toes, beaks, skin, eyelids, nostrils, tongues, necks, skulls, vertebrae, ribs, feathers, flying habits, eggs, and voices. But by the end of chapter 1, titled "Variation

Under Domestication," it's hard to imagine a better illustration of the mutability of species and of how selection can rapidly drive a startling degree of change and a dazzling diversity of forms, all from the starting point of the humble rock pigeon, known to city dwellers the world over. From this point of departure, it becomes almost impossible *not* to see how the same thing is happening in nature: Across geological epochs and in front of our eyes today, *natural* selection filters the adapted from the maladapted, creating a remarkable degree of fit between species and their environments. The conclusion seems inescapable that "from so simple a beginning endless forms most beautiful and most wonderful have been, and are being, evolved."

Darwin's book was a masterpiece, but as we have seen from our brief historical tour, the individual components of the theory had precedents going back centuries or millennia, and at least two other people independently proposed more or less the same coherent account of biological evolution. What Darwin accomplished was an unprecedented degree of rigor and detail, which he then wove into a compelling narrative, propelling evolutionary theory forward and into public view.

The Deep Roots of Evolutionary Ideas in the Social Sciences

The history of evolutionary thinking in the realm of language, technology, and ideas—what we might broadly call culture—involves not only a different cast of characters but also a fundamentally different narrative structure than the story for biology. During the two hundred years leading up to Darwin, natural historians shared a common goal: understanding life. In contrast, efforts to understand language, economic growth, or scientific advancement were undertaken as largely independent efforts, in different corners of the scholarly world. The story of cultural evolution thus involves many parallel currents of knowledge, the commonalities of which were only recognized relatively late in the game.

For some aspects of culture—technology in particular—common experience allows us to reject the hypothesis that God created and fixed their characteristics for all time. In the eighteenth and early nineteenth centuries, people saw with their own eyes the evolution of steam engines, electric telegraphs, and photography. Other aspects of culture change much more slowly. Language is the prime example, and indeed it had been suggested that languages were created and fixed in time by divine intervention. In the 1700s, countering the claim of language constancy was Sir William Jones, whom we met in chapter 1. With some help from friends, Sir William systematically analyzed similarities and differences between languages, producing evidence that they likely traced their ancestry back in time to a common source. He focused specifically on what is now recognized as the Indo-European language family, which includes more than four hundred languages spoken in a geographic swath stretching from Europe to India. Effectively, language was shown to evolve. So compelling was the case for language descent with modification that proto-evolutionary biologists frequently drew on analogies from linguistics to make their case for a similar process applying to life itself.

For aspects of culture other than language, the main alternative to the evolution hypothesis has been human intentions, rather than God. God didn't invent the steam engine or the printing press, but humans did. Whereas biological evolution has been aptly described as "design without a designer," the most impressive elements of human culture appear to have been designed *with* a designer. When scholars have taken a close look at the products of human culture, however, individual designers are rarely as important as we imagine them to be. Rather, for everything from writing and storytelling to the economy and science itself, cultural evolution always involves a great many people and an immensely important role for trial and error, countless rounds of which result in the incremental accumulation of innovations. In short, culture seems to evolve via repeated cycles of variation generation, selection, and inheritance.

Ultimately, as we will see throughout this book, human intentions may have a key role to play *in* the evolutionary process, but they do not constitute an *alternative* explanation to evolution. Humans might do much of the selecting on cultural variants—just as they have done for domestic plants and animals over the past twelve thousand years—but people's selective intentions do not somehow make the process non-evolutionary. Poodles evolved from wolves (with human direction), just as wolves evolved from a common ancestor they shared with coyotes (without human direction). At the start of this chapter, we already saw the idea of (natural) selection reflected in William Whewell's description of how science advances. The same idea cropped up in many other corners of the scholarly world in the eighteenth and nineteenth centuries, especially notably in economics.

The writings of economist Adam Smith (1723–1790) contain some cogent evolutionary perspectives. Smith's *The Wealth of Nations* begins by describing division of labor in a pin factory: "One man draws out the wire, another straights it, a third cuts it, a fourth points it, a fifth grinds it at the top for receiving the head." How did this efficient system come to be? Smith argued that division of labor was "not originally the effect of any human wisdom"; rather, it evolved by trial and error. Selection for more efficient production methods occurs within and among companies, and the best results are inherited by future versions of the same companies, and by any new companies that enter the industry. This is economic evolution, described long before anyone would have given it that name. Referring specifically to Adam Smith, the evolutionary biologist Stephen Jay Gould went so far as to say: "Darwin may have cribbed the idea of natural selection from economics." Although neoclassical economics would go on to focus more on the *lack* of change—or evolution—that results from a balance between supply and demand, the pathways to reaching an equilibrium (e.g., the total sales and price of a given product) are fundamentally evolutionary in nature.

The subject of economics bleeds subtly into the subject of technology, broadly defined to include everything from a stone hand axe or a metal pin to warships and wireless communications. If one person can build a boat in a few months, in what sense is the boat a product of evolution rather than a product of that person's intelligence? The philosopher David Hume, writing in 1779, questioned the ingenuity of the individual ship builder, asking if instead they were "a stupid mechanic, who imitated others, and copied an art, which, through a long

succession of ages, after multiplied trials, mistakes, corrections, delib-
erations, and controversies, had been gradually improving." In short,
there's no way the builder *could* build a workable ship if not for the
countless trials and errors of their predecessors from which to learn. Just
as no fish ever gave birth to a human, no human ever looked at a small
raft of logs and proceeded to build a 70-m-long warship with four masts
and one hundred guns. In Hume's words, we can once again see the
three key components of an evolutionary process: variation, differential
success, and inheritance. More than two centuries later, our intuition
still often leads us to imagine that technological innovations spring
forth fully formed from the minds of geniuses, when in fact they typically represent one incremental step on an evolutionary trajectory.

We can draw several important lessons from these early evolutionary
ideas originating outside of biology: First, biological evolution is just
one manifestation of a more general evolutionary process involving
variation generation, differential success, and inheritance. Second, similar ideas arose in numerous corners of the scholarly world. Finally, there
have been mutual influences among linguists, economists, biologists,
and scholars of technology for at least the past three centuries. While
contemporary writing on evolution outside of biology is dominated by
analogies flowing *from* biology, this is, in fact, a reversal of the dominant
flow prior to the mid-1800s, which was *into* biology from the social
sciences.

Biology Since Darwin

The theory of evolution by natural selection was not immediately
greeted with unanimous acceptance. For some, the evidence that Darwin presented made the truth of the theory so obvious that the real
mystery was why it hadn't come to light earlier. At the same time, doubts
about evolution by natural selection arose not only in religious circles
(predictably), but also in scientific circles, given some crucial remaining
uncertainties. Most important was the mechanism of inheritance. It was
plain to see that animal and plant offspring resembled their parents, but
how exactly was this achieved? The typical assumption at the time was

"blending inheritance," whereby the traits of offspring were intermediate to those of their parents. But blending presents a serious problem for the theory. If a pair of parents who are short and tall, or stocky and slender, always give rise to children who are of average height and build, the variation needed for natural selection to act on would quickly disappear. Everyone would be average everything.

With blending inheritance in mind, Darwin proposed a theory by which small particles called gemmules were produced throughout an organism's body, passed to sperm and eggs (gametes), and thereby inherited by the next generation. In later editions of *Origin*, Darwin also accepted a role for the inheritance of acquired characteristics (Lamarck's idea), with gemmules being influenced by the environment in which the organism lived, thus restoring some variation lost by blending. Unlike most of Darwin's ideas, this one would not survive scientific selection. And without a convincing theory to explain where variation came from and how it was inherited, the theory of evolution by natural selection would remain frustratingly incomplete. Fortunately, help was coming from some experiments already underway.

In the 1850s and 1860s, unbeknownst to Darwin, an Augustinian friar named Gregor Mendel was conducting experimental matings between different kinds of plants, ultimately leading him to discover the genetic mechanism of inheritance. Mendel focused mostly on the characteristics of the flowers, seed pods, and seeds of common pea plants. For example, pea plants make seeds with either a smooth surface or a wrinkly surface, and matings between them produced some confusing results. Sometimes a mating between parents with different kinds of seeds (smooth and wrinkly) produced offspring that all had the same kind of seed (smooth; see fig. 2.1, top panel). Sometimes a mating of two parents with the same seed type produced a portion of offspring with seeds unlike *either* of the parents (fig. 2.1, bottom panel). To make sense of these results, Mendel posited "factors" that came in pairs in each plant, one of which came from its mother and the other from its father. Together, these factors determined seed coat texture. We now call these factors genes, with each member of a pair called an allele. For what we might call the seed-coat-texture gene, each plant has two alleles, with

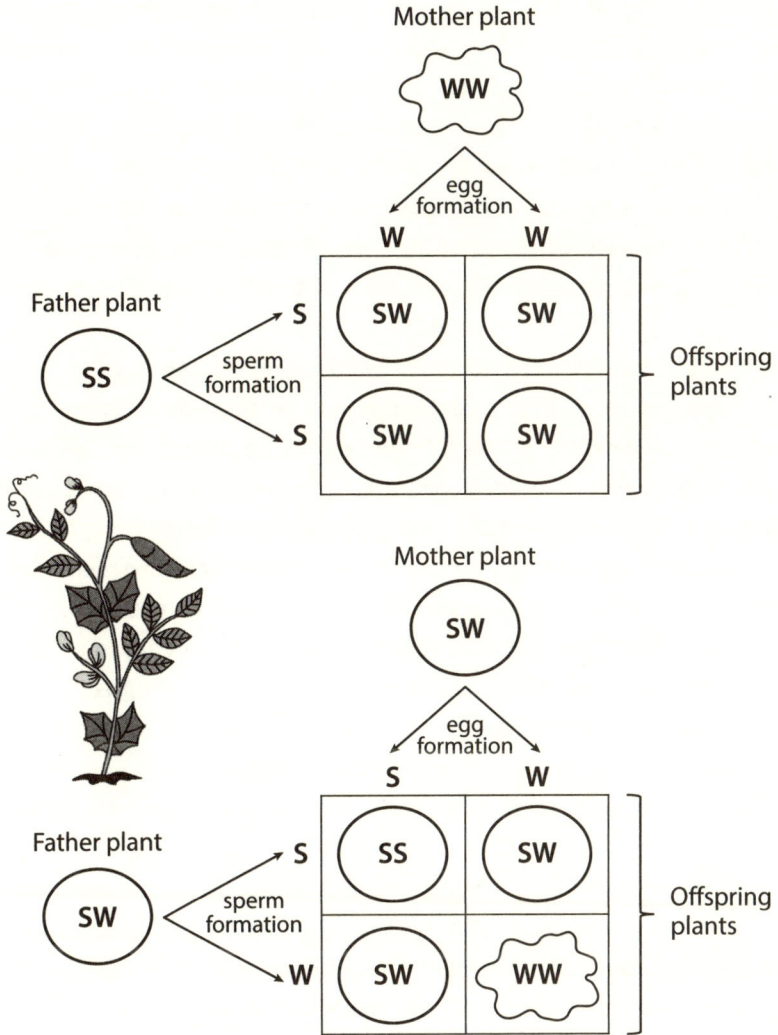

FIGURE 2.1. Results of some of Gregor Mendel's mating experiments, showing how parents with different phenotypes (smooth and wrinkly) can have offspring all with the same phenotype (*top*), and how parents with the same phenotype can have variable offspring (*bottom*). (Matings of two SS or WW parents give only SS and WW offspring, respectively.)

S "dominant" over W, so that SS and SW plants both have smooth seeds, while only WW plants have wrinkly seeds. In the end, from a complex set of results, Mendel offered a fairly simple explanation. His results went largely unnoticed and entirely unappreciated during his lifetime, having been rediscovered only at the turn of the twentieth century.

There are two key evolutionary lessons from Mendel's model of genetic inheritance. First, variation at the genetic level is preserved across generations because the process does not involve blending, but rather the combination of discrete "particles" (alleles). Biologists call this *particulate* inheritance. The second key lesson, emerging from analyses of multiple genes and traits together, is that the outcome of inheritance is largely independent across different genes. In other words, what happens with one gene for seed coat texture doesn't influence what happens with a second seed-coat-texture gene, or what happens with a gene for seed color (e.g., green versus yellow). A plant can have seeds that are smooth-yellow, smooth-green, wrinkly-yellow, or wrinkly-green. If we start to imagine the hundreds or thousands of genes in a genome, the number of possible combinations grows very large indeed. So, even though genes come in discrete packets, their combined influence can nonetheless produce a continuous distribution of traits, which is commonly observed for things like human height (short, tall, everything in between).

While Mendel's experiments and theory explained how variation was maintained across generations, they didn't explain where novel variation came from in the first place. No amount of creative mating among pea plants can produce a clover plant in the next generation. Since well before Mendel, scholars talked of mutation as the process by which one organism could give rise to quite a different organism, and for a time "mutationism" was seen as an alternative to evolution by natural selection, especially if mutations could create big differences. Smaller differences might be harder to detect, but by creating large numbers of offspring from large numbers of matings, biologists were able to detect rare genetic mutants. The pioneering geneticist Thomas Hunt Morgan (1866–1945) used chemicals and radiation to create mutant fruit flies, with the hopes of finding new mutations that would be heritable. A first success came in 1910, when a white-eyed fly was found in a population of otherwise red-eyed flies. Many subsequent studies showed how novel, heritable variants could arise via random errors of some sort. One mutation won't create a clover from a pea, but thousands or millions of mutations, accumulated and selected over thousands or millions of generations, can do the job. (Peas and clovers are in the same family of plants.)

Having achieved a basic understanding of inheritance, variation generation (mutation), and natural selection, all the puzzle pieces were in place for a coherent, synthetic picture of biological evolution to emerge in the early twentieth century. Many new observations, experiments, and concepts contributed to what later became known as the "modern synthesis." But it was the expression of evolutionary theory in mathematical form that was perhaps its most distinguishing and influential feature. Mostly from the 1920s onward, mathematically minded biologists such as R. A. Fisher (1890–1962), Sewall Wright (1889–1988), and J.B.S. Haldane (1892–1964) built many kinds of mathematical models, which remain the building blocks of evolutionary theory today. Some models focus on a single gene, while others assume an influence of many genes on traits such as the size of animals or plants. With a core set of models in place, researchers could predict how evolution might proceed in almost any imaginable scenario, depending on things like the strength of selection or the rate of generating new variants. Amazingly, the modern synthesis was achieved without anyone knowing what constituted a gene in biochemical terms, or how exactly those genes went on to influence traits. Figure 2.2 depicts some key aspects of the evolution of evolutionary biology between roughly 1850 and 1940.

Scientists eventually located the "instructions" for life in the nucleus of a cell, then in chromosomes that they could see under a microscope. Chromosomes were eventually found to be long, coiled-up molecules of deoxyribonucleic acid (DNA), expressing life's instructions using an alphabet of four "letters" known as nucleotides. In an organism like humans, chromosomes come in pairs, with one from each parent. In a process called meiosis, genes are swapped between the two chromosomes in a pair (this is why the texture and color genes for seeds are

Biological evolutionary theory circa:

Issue	Options (1850)		Options (1900)	Options (1940)
Earth age	Young	Old	**OLD**	**OLD**
Species' characters over time	Fixed	Modifiable	**MODIFIABLE**	**MODIFIABLE**
Species & adaptations: origin	Created by deity	Evolved (from other species)	**EVOLVED**	**EVOLVED**
Mechanism of adaptation	Inheritance of acquired characters	Natural selection	**NATURAL SELECTION**	**NATURAL SELECTION**
Inheritance	**BLENDING**		Blending / Particulate	**PARTICULATE**

Selection pressure via: Evidence (e.g., fossils), Social factors (e.g., religion)

Novel variants via: discovery (e.g., pea experiments), influx of results from other disciplines (e.g., geology, biochemistry)

FIGURE 2.2. The evolution of evolutionary biology from 1850 to 1940. Each row is a component or "trait" of evolutionary theory. Variation is represented by multiple options for a given component, with selection among them, sometimes producing a consensus shown in bold capitals.

inherited independently), and sperm and eggs are then produced with a single copy of each chromosome (otherwise offspring would have double the DNA of their parents). The "central dogma" of molecular biology specified a one-way flow of information: DNA provides the information to make RNA, which provides the information to make proteins, which do the work of building an organism, determining its size, shape, color, and so on.

The essential features of evolutionary theory developed by the 1950s were largely unchanged in the version I learned as a student in the 1990s and 2000s. As a single package, this combination of ideas is often referred to as neo-Darwinism, which is effectively a synonym of "modern synthesis." However, models of evolution built under the banner of neo-Darwinism involve some core assumptions that biologists have started to call into question. Questions about these assumptions are especially relevant for the prospects of a generalized evolutionary theory—that is, for the Second Science. First, biologists assumed for decades that the *only* means of inheritance across generations was the sequence of nucleotides in DNA. The authors of the modern synthesis largely had in mind animals that sexually reproduce. Such animals are produced from one egg plus one sperm, both derived from a cell line already present at birth (the germ line), and both of such tiny size compared to the adult animal that other possibilities for inheritance seemed limited. However, sexually reproducing animals make up but a tiny fraction of life on Earth.

The most numerous organisms on Earth are tiny single cells called prokaryotes, the most familiar of which are the Bacteria (a second group is called Archaea). Prokaryotes reproduce clonally, one cell splitting into two. The two daughter cells from a clonal split inherit essentially *everything* from the parent cell (proteins, cytoplasm, the cell membrane), not just DNA, so it is scarcely possible to even talk about "parents" or "offspring" when the offspring *is* the parent, pinched at the waist. Prokaryotes also routinely exchange DNA between species, violating two more assumptions of most evolutionary models: that inheritance flows only from parent to offspring vertically rather than between contemporaries horizontally, and that inheritance comes only from members of the same species. Moreover, neither clonal reproduction nor horizontal gene transfer between species is unique to prokaryotes: Both are found in a wide variety of animals and plants.

Even for sexually reproducing animals, along with a great many other species, researchers have found chemical alterations to DNA that do not affect the sequence of nucleotides but do affect the traits of organisms, and that are heritable. Among these *epigenetic* changes, the most well-known example involves a chemical tag that gets attached to DNA, influencing how a gene is expressed. Because these epigenetic changes can be acquired during the lifetime of an organism, enhancing its performance in the face of a particular environmental challenge, they conflict with not one but two assumptions of neo-Darwinism: First, they represent a form of inheritance other than that encoded in the nucleotide sequence of DNA. Second, heritable changes are not necessarily blind to their consequences, having been acquired during the lifetime of an organism, given an advantage they conferred.

One more type of nongenetic inheritance is important to mention: cultural inheritance. Some of the variation among animals in their behavior, such as their level of aggression or their mating tactics, is indeed genetically heritable. However, behaviors can also be inherited culturally—that is, by learning from other animals. This is patently obvious for humans, and there are many examples from nonhuman animals as well. Cultural transmission across generations—independent of any genetic change—has been observed for many animal behaviors, including songs in birds, and the use of tools in chimpanzees.

These insights about nongenetic inheritance have played a central role in prompting many biologists of the past twenty years or so to promote an "extended evolutionary synthesis" to replace the "modern" one. The argument for a new synthesis does not claim that the modern synthesis is *incorrect*, just that it is *incomplete*. DNA is not the *only* mechanism of inheritance, and not all changes to DNA are entirely random.

Cultural Evolution Since Darwin

In the first half of the nineteenth century, biologists drew on ideas and analogies from linguistics, economics, and technology, but there was no common point of contact for evolutionary thinkers in different disciplines. With the publication of *Origin*, the flow of ideas and analogies reversed direction, with the theory of evolution by natural selection becoming the central meeting place for evolutionary perspectives. In 1863 the novelist Samuel Butler wrote a newspaper article titled "Darwin Among the Machines," describing the descent of the many "species" of "mechanical life." Language scholars were quick to draw connections to Darwin, with Darwin himself writing in *Descent of Man* (1871) that "the survival or preservation of certain favoured words in the struggle for existence is natural selection." Note that he didn't say that the word survival was *like* natural selection; he said that it *is* natural selection. In 1898 the Norwegian-American economist Thorstein Veblen asked, in the title of an essay, "Why Is Economics Not an Evolutionary Science?" While many of the early links made to Darwin were somewhat vague and imprecise, the influence of *Origin* quickly extended in many directions.

In the study of human affairs, evolutionary thinking has had an impact in two very different ways. The first way was to consider humans as one species of animal like any other, attempting to make sense of our psychology and behavior in evolutionary terms. What forces of natural selection might have favored the evolution of human language, or cooperation, or our unusually large brains? Initially it was a struggle to convince people that humans had descended from nonhuman life forms, but with that fact established, the biological evolution of humans is of no greater significance to the Second Science than the application of evolutionary theory to any other species. More interesting to the Second Science are the extensions of the variation + differential success + inheritance recipe well beyond biology.

In a book titled *Darwin and the Humanities* (1909), the psychologist James Mark Baldwin summarized the first fifty years of Darwin's influence on a variety of topics, including psychology, the social sciences, logic, and religion. Baldwin himself had earlier coined the term *social heredity* to describe the many things a child inherits via social learning, not only from their biological parents, but from anyone. Social heredity is essentially a manifestation of what we now call cultural inheritance, and without it, people wouldn't be able to read or write or build a car, and they wouldn't know where to place eating utensils on a table or on which side of the road to drive. Scholars saw the trial-and-error nature of evolution by (natural) selection in the process of individual learning, in the advancement of scientific knowledge, and in the society-level benefits of social institutions. They also had a vision of how cultural change, such as learning how to live in a novel and possibly harsh environment, could feed back to influence genetic evolution (e.g., biological adaptations for living in the novel environment). This reciprocal interaction is now described as "gene-culture coevolution." In short, many of the core concepts in the field of study now called cultural evolution have been established and linked conceptually with the theory of evolution by natural selection for well over a century.

Before turning back to the roots of contemporary studies of cultural evolution, it is important to revisit one colossal misstep in the history of evolutionary thinking, already mentioned in chapter 1: the notion that evolution implies progress—from lower to higher forms, from simple to complex, from primitive to advanced, from worse to better. Leading the way in the nineteenth century was the English polymath Herbert Spencer (1820–1903), who proposed a universal *law of evolution*, in which life, human society, and the human mind were all on a pathway of progress toward perfection. Spencer coined the term *survival of the fittest* in an attempt to capture Darwin's theory in a nutshell, although Spencer's view of evolution deviated from Darwin's in assigning the inheritance of acquired characteristics a dominant (rather than minor) role. If a person worked hard to improve their intelligence and ethics, these would be passed on biologically to their descendants. A package of ideas tracing their ancestry back to Spencer's writings on evolutionary progress and perfection later became known as *social Darwinism*.

Social Darwinism has no precise definition, but generally speaking it is associated with the idea that humanity would benefit from permitting unconstrained competition among individuals or societies. Helping people in need would only serve to perpetuate whatever character traits caused them to be in need in the first place, thus contributing to the genetic deterioration of humanity. Taken to its egregious extreme, this ideology has been used to provide pseudo-scientific justification for racism, for dictating which people should be allowed to reproduce (eugenics), and ultimately for genocide. Up until the Second World War, a considerable number of respected scientists subscribed to these views (if not always the most extreme versions), although this number has since become tiny. Scientists have also rejected the notion of progress: Contemporary bacteria, fungi, plants, insects, and humans are equally adapted to their circumstances, all occupying twig tips on the tree of life, none lower nor higher than any other. Nonetheless, both the ideology of social Darwinism and attempts to borrow scientific credibility from Darwin persist. Fortunately, these dark roads of history did not completely derail efforts to explore legitimate scientific applications of evolutionary ideas—whether Darwin's or anyone else's—to a wide range of phenomena.

One area in which the application of evolutionary ideas has especially deep roots is economics. The subfield of evolutionary economics explicitly recognizes repeated cycles of innovation that disrupt any tendency toward reaching an economic steady state. Joseph Schumpeter (1883–1950) is credited as a key inspiration for this line of work, having popularized the term *creative destruction* to describe economic cycles of old being replaced with new. The modern field of evolutionary economics is often described as a sharp break from "orthodox" economics, which has a strong emphasis on predicting steady (equilibrium) states. In some models, "routines" of business firms are described as being like genes, which can mutate and recombine to produce innovations that constantly disrupt economic interactions. Although evolutionary economics has been defined as a counterpoint to orthodox or neoclassical economics—the dominant economic paradigm for over a century—the neoclassical approach is itself evolutionary in important ways.

Neoclassical economic models often begin by assuming that perfectly rational agents (people or firms) have perfect knowledge of market conditions and constraints. The models then predict what those agents—buyers, sellers, competitors—will do to maximize their profits or "utility." Conceptualizing it all as a game of sorts, models under the banner of "game theory" have been used to predict equilibrium strategies, such as the price of a given product or the rate of selling and buying. In a revealing turn of events, evolutionary biologists were quick to adopt game theory. If we consider biological reproduction as a form of profit and define games for particular biological scenarios, we can predict behavioral strategies—e.g., how an animal secures mates—that natural selection should favor. In short, even though these kinds of models focus on a final, unchanging state of the system (the equilibrium), instead of continual change, the pathway toward that final state can be unambiguously seen as evolutionary.

Many of the evolutionary models used in economics or in the study of culture have imported concepts from biology in a highly pragmatic way, using and adapting only those bits that help capture the essence of a particular scenario. Other efforts have aimed to zoom out from the details of any particular scenario to identify the core features of *any* evolutionary process—as I do in this book. In the 1970s and 1980s, evolutionary biologist Richard Dawkins speculated that life anywhere in the universe must evolve via a Darwinian process involving *replicators* of some sort: physical entities that make imperfect copies of themselves. Replicators, in turn, are embedded within "vehicles," or *interactors*. Dawkins called this *Universal Darwinism*. In earthly biology, genes fill the role of replicators, and organisms the role of interactor. For culture, the "meme" (a minimal unit of culture) was proposed as the replicator, and a brain as the interactor, but the analogies are limited. The word *meme* is now widely used to describe amusing bits of internet flotsam that spread rapidly, but in the study of cultural evolution, it is used only sparingly as an informal shorthand for an identifiable bit of culture. More generally, the effort to identify cultural counterparts for genes and whole organisms looks like a dead end. It is often impossible to delineate a cultural replicator as distinct from an interactor (try to do it for a cell phone or a language). As such, scientific selection has not favored the idea of a "Generalized Darwinism" based on the replicator-interactor dualism, even if the underlying sentiment of widely applicable evolutionary ideas has thrived.

Just as the study of biological evolution has been stimulated by mathematical models, so has the study of cultural evolution. Two books published in the 1980s, one by Robert Boyd and Peter Richerson, the other by Luigi Cavalli-Sforza and Marcus Feldman, applied a range of mathematical models to understanding cultural change. Borrowing heavily from population genetics, they attempted to make sense of why certain puzzling behaviors might spread (e.g., women choosing to have fewer children) and of how cultural evolution might be influenced by aspects of inheritance seemingly particular to culture, such as the tendency for people to copy prominent members of society. Over the past forty years, the study of cultural evolution has flourished and evolved.

The panoply of scientific subdisciplines now includes evolutionary anthropology, evolutionary archaeology, evolutionary economics, evolutionary epistemology, evolutionary ethics, evolutionary linguistics, evolutionary psychology, and evolutionary sociology, among others. With or without the label, evolutionary mechanisms have also been used to explain human history and temporal changes in technology, the arts, and political institutions.

Lessons from the Evolution of Evolutionary Thinking

The aim of this chapter was to provide the necessary historical backdrop to appreciate how we came to understand the processes underlying evolutionary change, which will be laid out in the next four chapters. We can see that evolutionary ideas have deep and diverse origins, having developed, aggregated, and disaggregated into and out of various packages over periods of decades, centuries, and even millennia. This is not a story of just a few towering figures doing all the work, nor is it just a story for biology. Evolutionary ideas have their deepest roots in biology, but for several centuries at least, attentive minds have also seen evolution at work in the realms of technology, language, and economics. Across all these fields, evolutionary ideas have been bubbling up since long before there was an evolutionary lexicon to describe them.

The history described here reveals how evolutionary science itself has evolved via repeated cycles of idea generation, selection or rejection of those ideas, and inheritance of the successful ones from one generation of thinkers to the next (e.g., fig. 2.2). At the same time, we can also see that alignment of a theory with the facts is not the only important selection pressure in the evolution of science. In the long term, assessing the validity of ideas based on their degree of agreement with the facts is what defines science, but along the way, many other factors can impinge on people's assessments. It is reasonable to suppose that widespread acceptance of evolutionary ideas would have occurred much earlier if not for opposition motivated by religious beliefs.

The deep and diverse roots of evolutionary ideas send a strong signal that the use of Darwin as a namesake for this complex body of knowledge has outlived its utility. At present, Darwin's name is used as shorthand for the modern evolutionary synthesis even though Darwin had no knowledge of the mechanistic details in the modern synthesis. Most synthetic treatments of cultural evolution are similarly described as Darwinian. From the view of the Second Science, neo-Darwinian evolution does not include *all* evolutionary processes and phenomena, and so tying the project of generalized evolutionary theory to the question of whether evolution is specifically *Darwinian* is a major and unnecessary distraction: the Darwinian Distraction. Associating a single name with all we have learned about evolution also gives a misleading sense of how this episode of cultural evolution actually played out.

In the end, neither biological evolution nor cultural evolution fits the Darwinian label perfectly, but both fields of study fit perfectly within what we can now call the Second Science. As such, the time has come to dive into a more formal declaration of the Second Science, and what we know about how variation is generated, inherited, selected, and moved around in evolutionary systems of all kinds. How and why does everything evolve?

3

Evolution at Many Levels

Grandparents love to subject their grandchildren to armchair genetic analysis. "She has her mother's eyes, but her father's smile." "His athletic ability comes from Dad; his brains from Mom." And they always seem to find one or more of junior's attributes that managed to skip a generation or two: "Great Grandpa Charlie is the only other person in the family who had such bushy eyebrows." In executing their analysis, Grandma and Grandpa (and uncles and aunts and friends) are applying—whether consciously or not—a mental model of how heredity works. According to their model, the shape of your eyes, or of your smile, comes to you as a whole package, either in mom's egg or in dad's sperm, or sometimes directly (and mysteriously) from a distant ancestor.

In our everyday lives, each of us is constantly making use of mental models of one sort or another. These models are essential if we are to have any hope of extracting meaning and understanding from a potentially overwhelming set of inputs to our brains. Mental models help us decide what to do on the weekend, by allowing us to assess the benefits and risks of drinking alcohol, of driving a car, or of getting a good night's sleep. A mental model of how the economy works helps us understand an economic downturn, perhaps prompting us to lay blame with the government or big business. We can grapple with the exquisite complexity and uniqueness of each human being by using a mental model of how genes make a person. In all cases, as we gain knowledge and experience, we update our mental models.

45

The job of a scientist is, in essence, to formalize the processes of building and testing mental models. We usually start by expressing them with as much precision as possible, either verbally or using math, at which point we can drop the word "mental" and just call them models. Scientific models represent hypotheses about how the world works, and they are tested by asking how well they align with reality. Has government (in)action caused past economic downturns? Can athletic ability be attributed solely to one parent or the other? The match with reality is rarely perfect, and the nature of imperfections leads us to develop new models that will hopefully perform better in the next round of tests. Science evolves.

According to contemporary models of heredity—that is, biological inheritance—individual gene copies (alleles) can indeed be attributed to one parent or the other, but traits like athletic ability are typically influenced by many genes (not just one), coming from both mom and dad. And genes can't actually skip generations. Unusual gene *combinations* might occur in relatives separated by multiple generations, but even this doesn't necessarily imply a causal chain of events. That is, Great Grandpa Charlie's bushy eyebrows might have nothing at all to do with yours.[1]

The point here is not to highlight the faults in Grandma and Grandpa's mental model of heredity. Their model does capture some important aspects of reality, and you may indeed have gotten a bushy-eyebrow allele from Great Grandpa Charlie. Ultimately, all our mental models are oversimplified, at least outside our narrow areas of expertise. The point is that the essence of science is nothing more than a deliberate and formal approach to what all of us are doing all the time: developing and testing models that help us make sense of the world. In the remaining chapters of the book, I will often talk about building models—of everything from cultural practices and technologies to business firms and bacteria. What I mean by building evolutionary models is simply the act of thinking carefully about what might cause any of these things to change over time and how diversity is produced.

All models involve a certain degree of abstraction or generalization. Take maps as an example. A map is essentially a model of a given geographic area, and the fact that it represents features on the ground in an abstract or simplified way is precisely what makes the map useful. Otherwise, it wouldn't fit it in your hand or on a screen. On a digital map (e.g., in the default view in Google Maps), zooming in to the smallest possible area provides the finest level of detail, and so the least abstraction. At this scale, you can see shapes and addresses of buildings, and the names of the narrowest streets. If we zoom out to the scale of a whole city, a greater degree of abstraction allows us to see the big picture (literally and figuratively), without too many superfluous details: Buildings and the narrowest streets have disappeared, and we only see urban areas (gray), large roads (white), highways (yellow), and nonurban areas (green). At the scale of a large country, the only roads remaining are the highways, along with state or provincial borders, coastlines, and broad land cover types (e.g., forests and fields as different shades of green). The key lesson here is that abstraction allows a model to apply more broadly than it otherwise could, with the trade-off that it sacrifices some local details that can be incorporated only with a more narrowly applicable model.

Extending this lesson beyond maps, consider how we might choose to build a model of the evolution of the English language. A detailed model might refer to the specific influences of the Norman invasion of England, the writings of William Shakespeare, and the King James Bible. Such a model would provide rich insights into English specifically, but not much else. If, instead, we focus on more abstract concepts like borrowing from other languages, or changes in word length, and how they relate to semantics (meaning) and syntax (how words and phrases are arranged), we can zoom out and build models that apply to any and all languages. The key step in going from English specifically to language in general was to identify features that are common to all languages, not specific to any one or a few. The general model can still be used to understand how Shakespeare fits into the picture, but doing so requires us to use terms and concepts that allow formal comparisons with other languages.

The Second Science takes the process of abstraction and generalization a step further, identifying the key features that unite all evolutionary systems, from language and technology to viruses and ecosystems. The Second Science is a large-scale map of the evolutionary world. The role of this chapter is to lay down the broad outlines of the Second Science's core model, and to elucidate the key concepts that make it work.

Evolutionary Systems

Building on the initial description given in chapter 1, we can now make things more precise by describing four core components of an evolutionary system. Three of these components are both necessary and sufficient to define an evolutionary system. The fourth is a ubiquitous and important, if not absolutely necessary, feature of most evolutionary systems.

First, *in a population of some kind of entity, there is variation among the entities in their characteristics.* The entities might be tools, communication devices, or animals. Their relevant characteristics could include their shapes, versatility, or speed.

Second, *the characteristics are at least partly inherited by future versions of the entities.* Each new version of a tool, device, or animal has ancestors, from which it has retained at least some characteristics.

Third, *entities have differential success.* Sometimes success is unrelated to an entity's characteristics: Between two similar types of stone tool, gray and brown, the gray type might become more widespread just because the few craftspeople who made the brown ones died from disease, not because gray tools were better in any way. In this case, the direction of evolution is random with respect to characteristics—what we call *drift*. It could just as easily have been the makers of gray tools who died. When success or failure is causally influenced by one or more characteristics, for example, a stone tool is preferred and so becomes more prevalent because it is sharper, we say there is *selection*.

When the three conditions listed—variation, inheritance, differential success—are met, the distribution of characteristics (traits) in a population will necessarily evolve. For example, stone tools get sharper, telephones get more versatile, and birds get more aerodynamic. Repeated rounds of the three steps can, over the long term, produce astonishing degrees of fit between an entity's characteristics and the environmental challenges it faces. We call this *adaptation*: Knives are adapted for cutting, telephones for communication, and wings for flying. With selection for different characteristics in different subpopulations, evolution can also create diversity at a higher level. Bread knives and hunting knives, landline phones and cell phones, finches and sparrows—all these pairs evolved when variation within a type became variation between types. While in most cases the three components—variation, inheritance, and differential success—are fairly obvious for anyone to see, linking them to adaption and higher-level diversity is much less obvious.

In theory, these are the only three ingredients necessary to define an evolutionary system. In almost all real evolutionary systems, however, entities are distributed among multiple, semi-independent clusters of some kind, forcing us to consider one additional process. The previous paragraph already referred to "subpopulations," which is one kind of such clustering. The human cultures in different cities, countries, and continents undergo their own evolutionary dynamics, just like finches living on different islands. Whether we want to model evolution across the full set of subpopulations (e.g., countries or islands) or in just one, the additional process we need to consider is movement. Ideas, technologies, and individual animals will move from one place to another, influencing evolution of the whole system, and its component parts, in important ways. The cultures of England and France are each on their own evolutionary trajectories, but they are not entirely independent of one another. Movement is a fourth component needed in our core Second Science model.

The Evolutionary Soundboard

The definition of evolutionary systems just given indicates that our core Second Science model requires four components: variation generation, inheritance, differential success, and movement. Experts agree on this much, but they disagree on whether the nature of these four components is similar enough across cases as diverse as languages, businesses, and biological populations to be captured in one coherent core model.

THE EVOLUTIONARY SOUNDBOARD

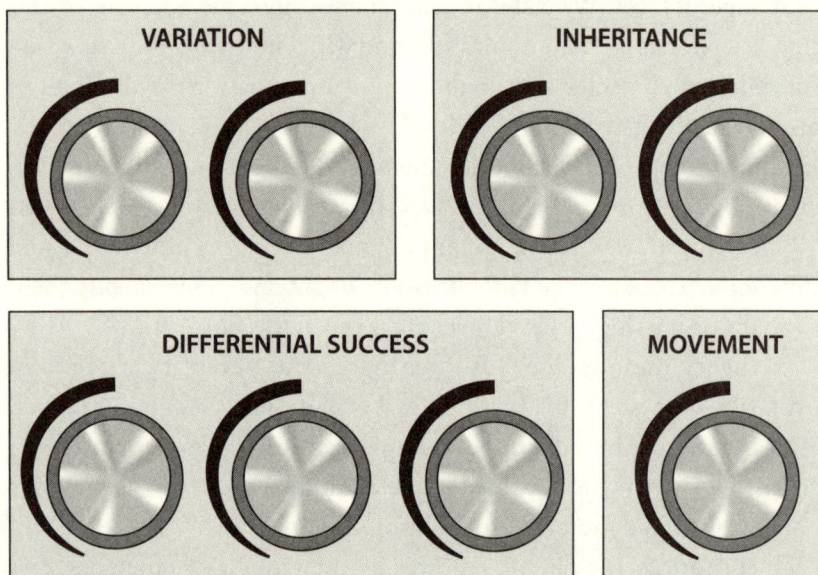

FIGURE 3.1. The Evolutionary Soundboard—simplified version. The dynamics of all evolutionary systems can be understood as the outcome of tuning these dials, which will be defined and applied throughout the book.

Here I propose the Evolutionary Soundboard (fig. 3.1) as a metaphor representing a core model that applies to all evolutionary systems in a coherent way.[2] For each of the four components—variation generation, inheritance, differential success, and movement—one or more dials can be tuned to capture that component's essential features. On the surface, it might appear as if moving from culture to biology, or vice versa, would require the addition or deletion of entire dials. However, by keeping an unwavering focus on a specific level of abstraction (all evolutionary systems), rather than attempting to force concepts (e.g., genes) from one discipline into another, the dials can be defined to apply universally.

At this point, these soundboard dials might sound a bit vague and mysterious. Chapters 4–6 will explain their underpinnings in detail, but for now, a brief preview of one set of the dials—those for inheritance—can serve to illustrate the utility of the Evolutionary Soundboard.

For a biologist, inheritance seems fairly simple: Genes come from your parents. Some species reproduce asexually (offspring have one parent), while others reproduce sexually (offspring have two parents), so in the biological version of the soundboard, we might imagine that one of the inheritance dials has just two settings: 1 or 2 parents. However, for horizontal gene transfer or for cultural evolution, traits can be inherited from many different sources, such that a dial with restricted settings (1 or 2) doesn't work. Instead, a dial that goes from 1 to some arbitrarily large number is needed to accommodate the full range of possibilities. Turn the dial to 1 for an asexual genetic parent, or if religious beliefs were inherited from just one person; turn it to 2 for sexual parents, or to 1000 if that's how many people influence your musical tastes.

The second inheritance dial captures the fact that not all traits are inherited from parents. The words *parent* and *offspring* are defined such that parents must come before offspring. So in biology, the fact that inheritance is vertical is typically just a background assumption, rather than something we need explicitly on the soundboard. But for bacteria exchanging genes, or for people developing language or technology, traits are often inherited horizontally from contemporaries rather than vertically from antecedents. Contemporaries might even be younger than the inheritor: An older bacterial cell can receive genes from a younger one, and a parent can be influenced culturally by their children. This requires a second dial in which we define the degree of vertical versus horizontal inheritance. A maximum value would mean that inheritance comes only from antecedents—the typical assumption in biology. A value of zero, the minimum, indicates only contemporary influences— a teenager's fashion choices come to mind.

In sum, it would not work to use the standard biological version of the soundboard to capture cultural evolution, but just a slight shift in perspective allows us to define two inheritance dials that capture the full range of possibilities, whether cultural, biological, or anything else. The range of levels to which each dial gets tuned overlaps between cultural and biological systems. What seemed like a situation calling for binary distinctions, or for adding or deleting entire dials depending on the situation, ultimately was captured by two universally applicable dials that can be tuned appropriately to characterize any and all evolutionary systems.

Throughout the rest of the book, the Evolutionary Soundboard will be a central organizing tool. Before digging into each of the dials on the soundboard in subsequent chapters, we first need to unpack some of the key concepts used to define evolutionary systems, specifically: *population, entity,* and *success*. As we shall see, these concepts require highly flexible definitions, depending on the specific evolutionary situation. Evolution happens simultaneously at many different scales of space and time, and at many hierarchical levels. Building models to characterize the dynamics of any given evolutionary system requires pinning down all these concepts, but there is rarely one correct way to do so.

Populations, Entities, and Success

To probe these key concepts, let's begin by considering a hypothetical experiment using slime molds, which, as the name suggests, often grow as a slimy film over a given surface. While the cells in most plants, animals, and fungi have one nucleus per cell, slime molds grow mostly without forming new cells, but just by increasing the size of one cell, which may contain many nuclei (where the chromosomes are housed). Our hypothetical experiment begins with one-third of each lab plate occupied by one of three types of slime mold: dark gray, light gray, or white (fig. 3.2, top panel). A selection pressure in the form of heating is applied to one batch of plates, while a second batch of plates (the "controls," not shown) is kept at the initial temperature. Compared to the control plates (where we observed no systematic changes), heating favors the growth of dark gray at the expense of light gray, but there is no change in the number of nuclei for either. White shows neither increased nor decreased size, but a 50% increase in the number of nuclei.

In biology, evolutionary success is often measured (at least in a rough sense) in terms of new copies of genes produced across generations. For the slime molds, we might therefore say that success was greatest for white. But the dark gray type was more successful at producing more of itself in terms of areal coverage and presumably mass, which might

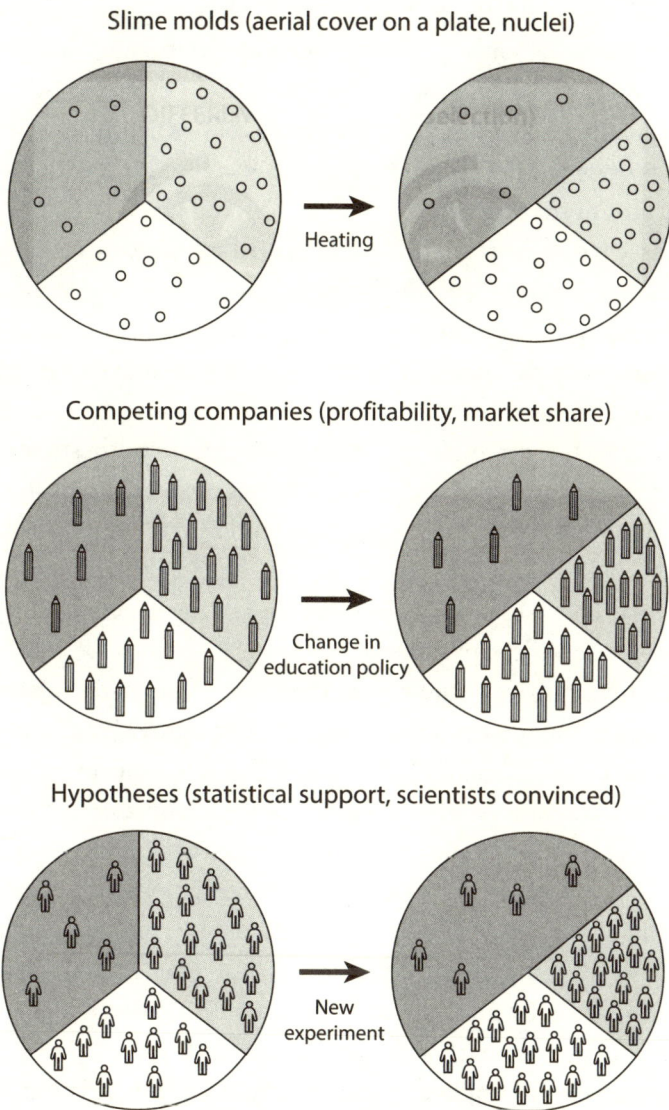

Slime molds (aerial cover on a plate, nuclei)

Heating

Competing companies (profitability, market share)

Change in
education policy

Hypotheses (statistical support, scientists convinced)

New
experiment

FIGURE 3.2. Hypothetical evolution in slime molds growing on a plate, companies competing to sell pencils, or scientific hypotheses competing to explain evidence. Possible measures of success are in parentheses, the first represented by the size of the pie slice, the second by the number of items inside each slice: nuclei, pencils, or people.

ultimately lead to a plate covered *only* with dark gray: clearly an evolutionary success for dark gray. Both views are legitimate, and we could build a model of the system using one definition of success, the other, or both simultaneously. To describe success of this nature, evolutionary scientists often use the term *fitness*. For now, we can stick with the more generic idea of success.

The essence of the scenario just described for slime molds applies to any evolutionary system in which success can be measured in more than one way. Each color might represent a company in the business of selling pencils, with the size of a company's pie slice representing profitability, and the number of pencils its market share (fig. 3.2, middle panel). Or each color might represent a scientific hypothesis, for which slices of the pie are sized according to a measure of statistical support for the hypothesis, and the stick figures represent the number of scientists who believe the hypothesis is true (fig. 3.2, bottom panel). Selection in these cases could be applied via a change in education policy or via new scientific experiments.

In the long run, we might expect the different measures of success to align—for example, what scientists accept as the provisional truth ought to be determined ultimately by evidence. But in the short term, we have multiple valid indicators of success. For a time, scientists might have valid reasons for doubting the results of a new experiment, or they might hold tight to their favorite hypothesis (e.g., the white hypothesis), rallying support from colleagues when feeling threatened by contrary evidence. In short, different components of success might respond in contrasting ways to the same selection pressure, and even when they ultimately align, it can take time for evolution to catch up with selection.

There are also multiple options for defining the population and the entities that make up a population. We could consider a population of nuclei, pencils, or people, which presumably vary in characteristics that determine their fate in response to heat, education policy, or experimental evidence. We could consider a population of cells, companies, or hypotheses, with just one of each color. Some options are more abstract, but no less legitimate. For slime molds, for example, we could consider a population of bits of areal coverage (or mass), which aligns with the view of increased size as success. We might not be able to quantify areal cover by counting something, but we can easily estimate the quantity in units of square centimeters. Biologists often take pains to distinguish reproduction (a common proxy for evolutionary success) and growth (a characteristic of one organism), but in these and many other cases there is no clear line to be drawn between the two, even within biology. No matter how we look at it, in all these populations, differential success has led to evolution.[3]

A potentially puzzling situation occurs when we describe the evolution of what seems like a single entity, such as a company, a recipe, or a society, independently of the evolution happening in the broader population of companies, recipes, or societies. To understand how we can solve this puzzle, imagine a new company initially selling pencils that are 10 centimeters long—short for a pencil, but much appreciated by certain customers (fig. 3.3, time step 0). Every year the management team considers possible changes to the business, such as increasing the range of pencil sizes. At step 1, they consider two options: In addition to keeping the 10-cm pencil, they can add either a 5- or a 15-cm version. They go with the 5 but quickly learn that there is no market for this, so at step 2 they consider returning to selling only the 10, or also adding a 15. Longer pencils prove popular, but eventually they overshoot by going all the way to 25 (steps 3 and 4), after which they return to selling pencils in the 10–20 range (step 5). There was only ever one company with one implemented strategy, but there were repeated rounds of generating variation (*possible* strategies) and selecting among them, together leading to the evolution of the company. More generally, even with just one configuration of a business, family recipe, or society at a given moment in time, there can be a sizable population of alternatives. These alternatives exist in the form of ideas, including both memories of past configurations and imagined future configurations.

We can also imagine a comparable scenario in the realm of biology. Picture a single bacterial cell living in a tiny container with a food supply that is only enough to support the one cell. Every day, the bacterial cell splits in two. One daughter cell is exactly the same as the mother cell, while the other has mutated to be a bit different. Whichever daughter

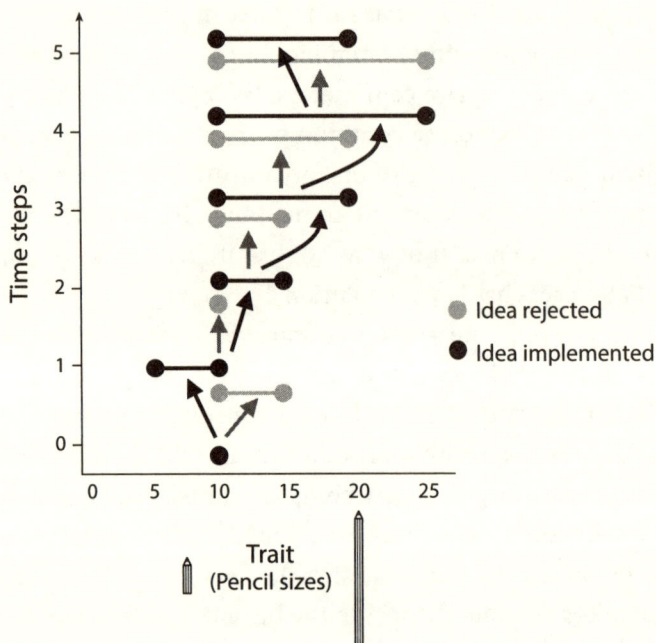

FIGURE 3.3. Evolution of pencil sizes sold by one company. Each horizontal line represents an idea for a range of pencil sizes to sell.

cell is not as good at taking up food immediately dies, since there is only room enough for one full-grown cell. Over time, we expect the one cell to get better and better at taking up food. In evolutionary terms, selection leads to the evolution of an increased rate of food uptake, even if there is only one cell present at a given time, apart from those fleeting moments after a split. This scenario is illustrated in figure 3.4.

These two examples—a company selling pencils and a bacterium in a tiny container—illustrate the flexibility of the concept of a population. Evolution requires variation among entities, but there need not always be more than one entity at a given time, and not all entities need to go beyond ideas or ephemeral failures. Evolution is the science of trial and error, so we do need repeated trials, but it is possible that most of the rejects never see the light of day. Single entities might also be involved in evolution at a higher level. In the case of pencils, the company might be one of many, such that evolution can also be

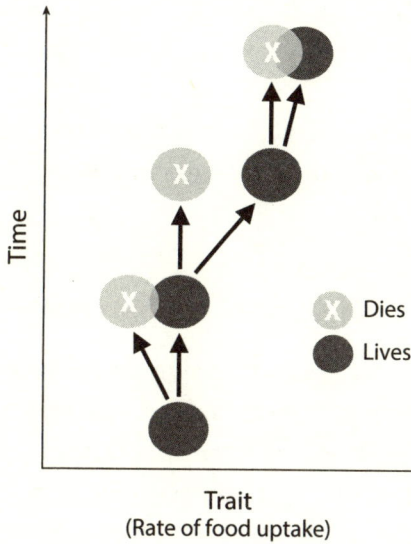

FIGURE 3.4. Evolution of the rate of food uptake in a container large enough for only one bacterial cell.

happening at the level of the entire pencil industry (see fig. 3.2, middle panel), which will no doubt feed back to influence evolution within a given company.

The concept of a population also implies that we can draw a clean distinction between entities that are members of the population and those that are not. In reality, there is often a considerable degree of arbitrariness in defining populations. Imagine two islands in the middle of the ocean. If the islands were small (let's say a few square kilometers) relative to the distance between them (let's say hundreds of kilometers), we might expect the plants and animals and human cultures on each island to be evolving independently, for the most part. As such, we might consider the entities on each island (trees, birds, cultural artifacts) as distinct populations. In contrast, if the islands were separated by a narrow straight just 100 meters wide, it is unlikely we would choose to consider them as separate. There is no magic size-distance combination that defines an unambiguous threshold. And even for small islands that are far apart (e.g., some of the Hawaiian Islands), if we want to compare biodiversity or cultures at the global scale, we might choose to consider the entire set of islands as one population. Ultimately, when designing evolutionary studies or building models, decisions will depend on our existing knowledge of the scales at which evolutionary processes such as selection seem to vary, and also practicalities of what we are able to observe. There are many different legitimate ways to delineate populations as objects of study.

Evolution at Multiple Levels

As the examples so far illustrate, evolution can be happening at multiple levels simultaneously, involving genomes or biomass on a plate of slime molds, single companies or entire industries, and human cultures within one island, across an archipelago, or even the entire planet.[4] Depending on how we delineate an evolutionary system, new variants might be ideas, prototypes, products, companies, genes, genomes, traits, or species, to name just a handful of possibilities. When applying the Evolutionary Soundboard, we need to be precise in defining the level at which we consider an individual entity to exist, how to recognize when a new type has been produced, and what constitutes the population of interest.

No matter how you choose to delineate your evolutionary system of interest—let's say it's cell phones—there will be some things that your model just needs to take as a given, such as the fact that new ideas emerge at particular times. Indeed, *all* models must make simplifying assumptions of this nature. When Wallace and Darwin presented their verbal model of biological evolution, there was little knowledge of how variation generation and inheritance worked. They just took as a given that these things happened somehow. Even at the finest level of detail, we need simplifying assumptions. Physicists, for example, must take as a given the mass and charge of fundamental particles. In a model of atomic bonds, physicists don't attempt to explain *why* electrons have their particular mass and charge, they just accept the measured values and move on. The key point is that we can build useful models at one level of analysis (e.g., technological evolution) without a complete accounting of what's happening at all other levels (e.g., the neurobiology of the ideas behind technology).

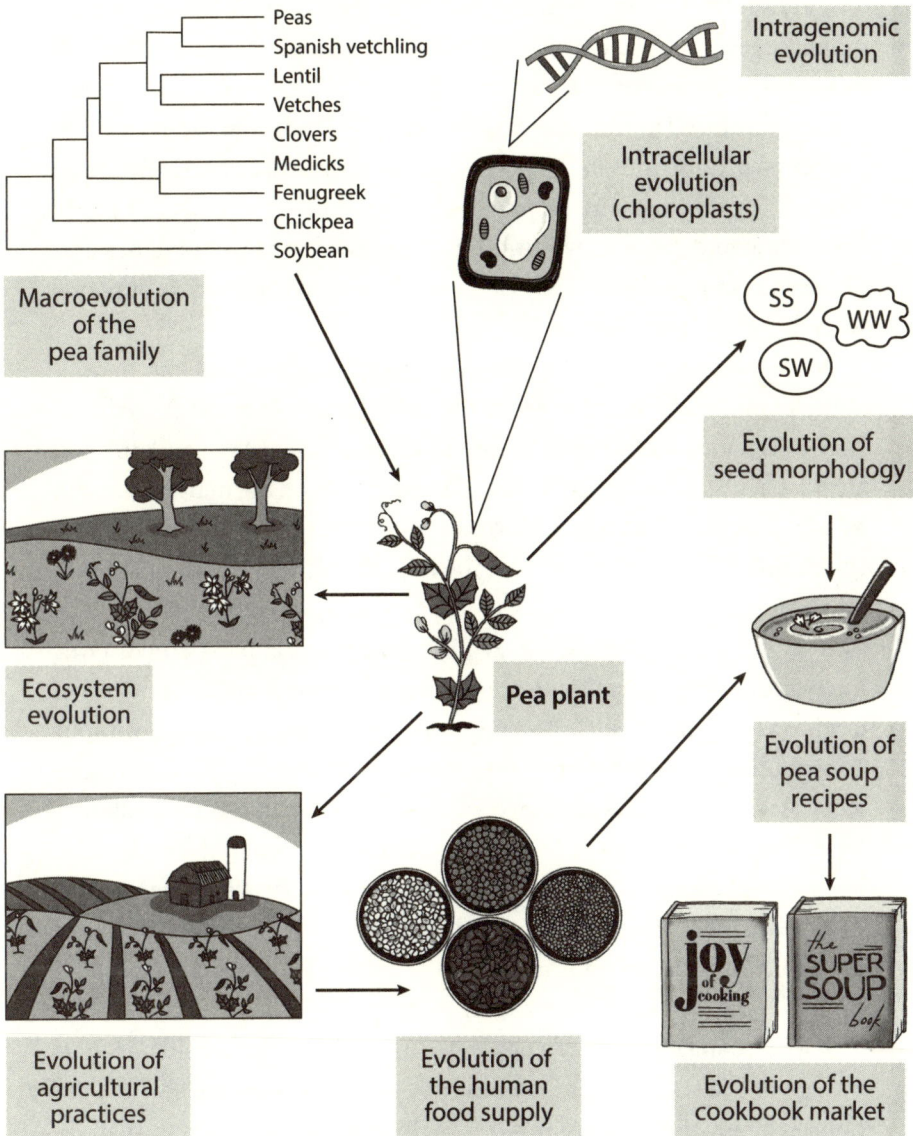

FIGURE 3.5. Evolution happens at many interacting levels simultaneously.

To begin understanding levels of analysis, let's revisit Mendel's pea plants (fig. 3.5). Smooth and wrinkly are two phenotypes at the level of individual plants. Without knowing precisely how these two kinds of seeds are made, we posit a hypothetical gene with two alleles, S and W. Plants with genotypes SS or SW make smooth seeds; WW plants make wrinkly seeds. To analyze evolution at the level of populations of individual organisms, biologists frequently build models in which the outputs are the proportions of individuals with different genotypes in the next generation (in this case SS, SW, or WW), and the inputs are the genotype proportions in the present generation and any selective (dis) advantage of having a given phenotype (smooth or wrinkly). A study like this would capture evolution happening at the level of genes and genotypes in populations of pea plants, based on the success of plants with different seed textures.

More than a century after Mendel, biologists figured out exactly why some plants make wrinkly seeds instead of smooth ones: A dysfunctional enzyme causes seeds to first swell to unusual size, only to then shrink back and become wrinkly. Compared to the normal S allele, the W allele includes a large section of DNA that has been inserted in the middle. And it turns out that the new chunk of DNA is most likely what biologists call a transposon. Transposons are often described as "selfish" or "parasitic," since they replicate and insert copies of themselves into various parts of a genome, sometimes compromising the function of a gene (as with the W allele) or simply making the genome larger without contributing to the function of the organism. This means that behind the story of wrinkly seeds is another evolutionary story at an entirely different level. The W allele was a novel variant when it first appeared, but its cause was the evolutionary success of a transposon, which itself can be thought of as a variant in the population of transposons. In fact, the majority of the DNA in plant and animal genomes appears to have accumulated partly via the proliferation of selfish chunks of DNA.

If we peer inside a cell of a pea leaf, the S or W alleles would be found on one of the seven chromosomes (full of junk as they are), which are housed inside the nucleus. It is this membrane-bound nucleus that defines what biologists call eukaryotes—which include plants, animals, and fungi. (Prokaryotes, such as bacteria, have their DNA floating freely inside the cell.) In addition to the nucleus, plant cells have other membrane-bound organelles—the "organs" inside a cell—called mitochondria and chloroplasts. Mitochondria generate the energy used to do most of the important work inside a cell, chloroplasts house the machinery for photosynthesis, and both originated some two billion years ago as free-living bacteria that were swallowed up by a larger cell, with which they now form an intimate coalition. Essentially, mitochondria and chloroplasts are tiny bacteria living inside of every plant cell, and they both still have little genomes of their own, separate from the seven pea-plant chromosomes. What's important here is that these organelles are still subject to evolutionary dynamics at their own level (generation of variation, differential success, inheritance) even if their fates are intimately linked to those of the cells and organisms that house them.

Now imagine wild pea plants growing in a prairie with a dozen other plant species—grasses, lilies, asters, etc. In this context, the difference between smooth-seeded and wrinkly-seeded plants pales in comparison to the differences between species, which show a great diversity of sizes, shapes, and physiologies. A model of change (i.e., evolution) in this ecosystem will look almost identical to our model of pea genotypes, with the outputs being the proportional abundances of each species at some future time (pea, grass, lily, etc.) instead of the proportions of genotypes. The inputs will be the species' proportions at present and any selective (dis)advantage of being one species or another. At this level, without allowing for immigration from elsewhere, a new variant arises when one of the resident species diverges—speciates—into two new evolutionary lineages. Speciation is also the mechanism by which new variants (species) are produced in macroevolutionary models, which might describe the process of diversification over the past sixty million years from a single pea-family ancestor to the twenty thousand or so species in the family today.

Domesticated plants and animals—whether peas, potatoes, dogs, or cows—provide particularly compelling case studies in evolution, as they span both the biological and cultural worlds. Selection imposed by humans underlies the conspicuous evolution of larger seeds, fatter tubers, more docile behavior, or greater milk production in domesticated plants and animals, compared to their wild counterparts. Over centuries and millennia, farmers and agronomists have devised countless new agricultural practices. Selection among alternative practices, imposed by individual people or collectives, has driven the evolution of everything from fertilizers, irrigation, and pest control techniques to methods of soil preparation, seed planting, and crop rotation. The evolution of breeding techniques and farming practices sparked the Green Revolution of the twentieth century, which produced massive increases in crop yields globally, with peas at the modest end of the spectrum, having doubled over the past sixty years. At the same time, movement of people and their products and ideas across the planet, along with local selection, has caused rapid evolution of the food supply: Between 1961 and 2009, the number of countries reporting peas as a crop commodity increased almost twofold.

Techniques for converting raw agricultural products into breakfast, lunch, and dinner have also evolved via trial and error in billions of kitchens across the world. In the French-speaking part of Canada where I live, peas are the core ingredient in a beloved traditional staple: pea soup. The evolution of recipes presents a familiar setting in which to think further about evolution happening at different levels. The evolutionary story behind a recipe and the evolutionary story behind pea plants are very different in some ways, but essentially the same in the most important ways.

In my own home, the family favorite is not pea soup but potato-leek soup, which serves our purpose here for considering recipe evolution. The evolutionary history of potato-leek soup in my own home started with the recipe in the book *Joy of Cooking*. The recipe there mentions a synonym for potato-leek soup, potage Parmentier, tracing the evolutionary lineage back (at least) to the eighteenth-century Frenchman Antoine-Augustin Parmentier, a prominent potato promoter in Europe. When I first started using this recipe, *Joy of Cooking* instructed me to use 6 leeks and 1¼ pounds of potatoes. That's more of a leek soup (too sharp) than a potato soup (so smooth), so I gradually shifted the ratio to more like one or two leeks for each pound of potatoes. The original recipe also says to first cook the leeks over medium-low heat in butter "until very tender but not browned." With a little distraction, heat too high, cooked too long, one time there was substantial browning, which made for . . . a much tastier soup! My favorite soup recipe has evolved.

As with pea plants, this little story about soup shows new variants arising and evolution happening at multiple levels. First, brainstorming possible improvements involved new ideas that were only ever considered at the level of my mind: no change in the leek-to-potato ratio, small change, or large change? Behind all that deep thought was something physical happening at the level of neurons, but even the best current neuroscience could probably not take a causal inquiry of novelty at that level much further than pointing to a likely section of the brain where soupy things were happening. Perhaps in fifty or a hundred years we will be able to trace the origins of thoughts to novel variation at a physical, neurobiological level, but for now we can't say much more about this than Darwin could say about the physical basis for biological heredity in the nineteenth century.

Moving beyond my brain, some preselected ideas (e.g., a small decrease to the leek-potato ratio) were used to produce novel variants at the level of actual batches of soup that could be subject to dinner-table selection. Interestingly, the multiple variants that made it this far never existed as physical bowls of soup simultaneously. Rather, it is the *memories* of variant soups, past and present, that provide the fuel for each round of selection, in which the whole family now takes part. While my tinkering with this recipe has focused largely on ingredient ratios, cooking time, and temperature, many other components could be involved. The butter, the bouillon cubes, the cooking pot, the stovetop, the potatoes, and the leeks are all on their own evolutionary trajectories— different now than they were for Antoine-Augustin Parmentier, and different from how they will be a century from now. As one more level to consider, imagine that we managed to convince the authors of *Joy of Cooking* to modify their recipe. This would create a novel variant at the level of a much broader marketplace of cookbooks. At all these levels of analysis, novel variants arise in meaningful ways, can be inherited over time, and achieve differential success.

As examples of evolutionary systems, there's nothing special about pea plants, agricultural practices, and soup recipes. We can see multiple levels of analysis in all evolutionary situations. Languages can evolve at the levels of the alphabet (letters), words, grammar, and syntax, or within individuals, local communities, or a global diaspora. Novel ideas for political institutions arise within individual minds, committees, and governments, and those that come to fruition represent novel additions to the global population of realized options. The prolific inventor Thomas Edison made no fewer than five hundred sketches of telephones, which were no doubt preceded by many ideas that never made it onto paper, and hardly any of which were built and given a chance to compete with other designs.

In addition, different levels and different systems can interact in myriad ways. The evolution of transposons within the pea genome can have important consequences for the evolution of pea populations. If the chemical composition or climatic tolerance of domestic peas evolved, that might influence not only recipes for pea soup, but also the evolution of agricultural practices, and if the change is big enough, an entire regional economy. These innumerable interactions present a challenge: If everything affects everything, the prospect of understanding evolution in general might seem overwhelming. To meet this challenge, evolutionary science moves forward by identifying particular levels of interest in a given situation, and then essentially starting afresh in building an evolutionary model at that level, making simplifying assumptions about most everything else, including what's happening at other levels or in other systems. As we will see in the next three chapters, all such models can be thought of as involving the tuning of dials on the universal Evolutionary Soundboard.

The Pragmatic Second Scientist

Overall, from our consideration of evolutionary models and levels of analysis, we can glean the key lesson that there is no one correct or best level at which to study evolution. All are perfectly legitimate objects of study and analysis (even if it is more difficult to get research funding to study the evolution of soup recipes than to study the evolution of food plants). As we build models of how technologies, economies, cultures, or DNA evolve, it is always necessary to treat some underlying mechanics in an abstract way. This applies to models of idea generation that abstract away the underlying neurobiology of the brain, and it is true of models with DNA mutation that abstract away the physical effects of the various mutagens (e.g., radiation or chemical agents). A central goal of this book is to propose the components and the degrees of abstraction that can still capture the most important aspects evolutionary dynamics at any level in any evolutionary system. It is time to begin fleshing out the Evolutionary Soundboard.

PART II

The Second Science

4

Where Does New Stuff Come from, and How Is It Passed on Through Time?

I use Google dozens of times every day to search for information on . . . checking my history . . . fellow scientists, local road closures, definitions of words, pea agriculture, and how to make enchilada sauce. You probably do too. It takes a tiny fraction of a second for the results to pop up—not enough time to be perceived as a delay at all—and invariably we find what we're looking for near the top of the list.

If I google "pea plants," one of the first results is a page in the *Old Farmer's Almanac* (http://www.almanac.com). From what we know of Google's algorithm, this means that many other websites contain links that point to the almanac (these are called "backlinks"), and many of those other websites are themselves frequently linked to (i.e., they are considered to be of high quality). In more general terms, Google's algorithm for ranking search results is based, to a considerable degree, on the number and quality of backlinks. Because links are created by people who build websites, the algorithm effectively harnesses knowledge of human behavior to predict what results will best meet your needs. And it works incredibly well.

Implementing the Google algorithm requires data on the billions of links that connect hundreds of millions of websites. In essence, it requires a map of the internet—not a simple thing to create. You might think that the founders of Google, Larry Page and Sergey Brin, first decided to build an internet search engine and then realized they needed to map the internet. In fact, the order of operations was reversed. As a PhD student at Stanford University, Page first had the goal of mapping backlinks. The idea was that such a map might facilitate communication across websites or permit analysis of the structure of the internet. According to Page, he "had no thought of building a search engine": "The idea wasn't even on the radar." In the end, Google's global domination of internet search was a serendipitous outcome of a project with an entirely different aim. In the evolution of internet search engines, a new variant arose by accident.

In this chapter we will focus on the emergence of novel variation and the inheritance of traits, without which there can be no evolutionary change in technology, culture, or ecosystems. One barrier to generalizing evolutionary theory has been the idea that cultural and biological mechanisms for generating new variation are fundamentally different. In cultural evolution, new variants are said to be introduced by people with intent, because those variants are predicted to be useful in some specific way. In biology, the main source of novelty is mutations—changes to the sequence of nucleotides in DNA—which are said to be random and "blind" with respect to their consequences. This view suggests two distinct options: cultural variation that is directed and biological variation that is random.

The origin story of Google allows us to reject the assumption that all cultural variation is directed, as do many other cases of happenstance in the evolution of technology and culture. Alexander Fleming discovered the first antibiotic by accident when some of his laboratory plates of bacteria got contaminated by a fungus that turned out to be releasing something that killed the bacteria. Thomas Newcomen discovered a powerful means to condense steam and to greatly enhance the efficiency of steam engines, because of a cold-water leak. We could go on. At the same time, we find some striking evidence in biology undermining the assumption that biological variation is always generated randomly.

In the 2010s the precision and efficiency with which scientists can alter genomes took a major leap, when biochemists figured out how to co-opt a trick used by bacteria for defense against viruses. The technology is called CRISPR-Cas9, for which Emmanuelle Charpentier and Jennifer Doudna received the 2020 Nobel Prize in Chemistry. The world's attention is now rightly focused on the mindboggling opportunities, as well as the profound dangers and ethical quandaries, presented by this technology. But CRISPR-Cas9 caught my eye for an entirely different reason.

While the human immune system serves to fend off infectious bacteria (among other things), bacteria themselves have an immune system that functions to defend against potentially lethal viruses, known as phages. When a phage docks on the bacterial cell membrane, its DNA is inserted into the cell for replication and the production of new virus particles. In many bacteria and other prokaryotes, scientists have found an ability to incorporate short portions of viral DNA into regions of their genome called Clustered Regularly Interspaced Short Palindromic Repeats, or CRISPR for short. The DNA segments in the bacterial genome that are of viral origin are called spacers, which effectively allow the cell to remember the viruses it has been infected with. When the same virus (or a very similar one) returns, RNA molecules made partly from the spacer DNA can find a match in the viral DNA, allowing an enzyme to come along and cut the foreign DNA, stopping viral replication before it can start. One such enzyme is called Cas9, and the innovation of Charpentier and Doudna was to engineer a simplified version of the CRISPR-Cas9 system to permit cheap, easy, and precisely targeted cuts and edits to DNA.

As awe-inspiring as the technology is, more interesting to the Second Science is what is happening naturally with the bacteria. In chapter 2 we saw how the modern synthesis of evolutionary biology was challenged by aspects of bacterial biology, such as rampant horizontal gene transfer and clonal reproduction. It turns out that bacteria have some additional habits that give evolutionary biologists even worse headaches. In the case of CRISPR, an external challenge (infection by a virus) ultimately causes a change to the cell's genome that is passed on to daughter cells and that is beneficial in the face of that same challenge. This flies in the face of the central tenet of neo-Darwinism and the modern synthesis that genetic

mutations happen at random with respect to their potential consequences. My own training in biology involved ridiculing any suggestion
that evolution involved the inheritance of beneficial characteristics that
were acquired during the lifetime of an organism. This makes the biology
of CRISPR astonishing. Giraffes might not have inherited long necks
because their ancestors stretched to reach the treetops (as Lamarck en-
visioned), but bacteria sure seem to have inherited immune systems that
were stretched by their ancestors to deal with new viruses.

CRISPR is an especially blatant case of nonrandom mutation, but
there are many others. For example, in a range of situations, researchers
have found that the parts of the genome most important for an organism's function are also the parts least likely to change via DNA mutation.
In other words, cellular machinery protects the most important parts of
the genome from mutation, while permitting more mutations where the
potential negative effects are least. Changes to the diet of fruit flies can
induce an altered number of copies of certain genes that help build ribosomes (where proteins are made), and the altered genome is inherited by their offspring. In short, the deeper researchers dig into studies
of mutation, the more examples they find in which things are not as
random as we thought. And the more we learn about genetics, the more
easily genetic engineers can deliberately introduce specific mutations
into a wide range of species, including pigs, cows, fish, humans, and at
least thirty plant species.

The question of random versus directed variation is just one among
many situations we will encounter over the next three chapters in which
different evolutionary systems—*all* evolutionary systems—can be
compared without any need to divide them into distinct categories (e.g.,
biological and cultural). Rather, we can determine where each system
falls on a continuum of one sort or another, such as the rate of variation
generation or the strength of selection. Any categories we can think of,
such as asexual versus sexual species or biology versus culture, will over-
lap on one or more of these continuous axes of variation, which is one
of the essential reasons we need a generalized evolutionary science—
the Second Science—rather than a haphazard collection of semi-
independent areas of inquiry that share the word evolutionary. My goal

over the next three chapters is to identify a universal starter kit of variables that characterize the processes driving change in any evolutionary system. These will make up the dials on the Evolutionary Soundboard, but before we get there, we first need to dive into what lies behind the dials for variation generation and inheritance.

Variants Are Generated and Inherited in Many Ways

In one sense, variation and inheritance are simple: Current and future versions of ideas, recipes, genes, organisms, businesses, and ecosystems are imperfect copies of previous versions.[1] The fact that future versions are *copies* of past ones (at least in a loose sense) means there is inheritance; the fact that copies are *imperfect* means there is variation. As described in chapter 3, what counts as a new variant will depend on the level at which we view a given evolutionary system. At the most granular level, the explanation for how a new variant is produced lies outside of evolutionary science, for example in the chemistry of the brain (for ideas) or in the interactions of chemicals with other chemicals or with radiation (for DNA mutations). In other cases, the new entities might themselves be products of evolution at a lower level: The iPhone evolved considerably within Apple before being released as a new variant of cell phone, subject to selection in the marketplace.

New variants can be generated in myriad ways. At least two broad categorizations that apply across the full sweep of evolutionary systems can help us wrap our heads around the diversity of specific ways that cultural or biological variation can be generated. One is the distinction between random and directed variation, discussed in the previous section. The other distinguishes new variants that are produced by *tinkering* from those that are produced by *combining*.

Tinkering with an existing design of a recipe, a computer, or a genome can happen via an altered ratio of ingredients, packing more and smaller transistors onto a computer chip, or a random DNA copying error, usually resulting in fairly small overall changes. Much larger changes can happen via *combination* of multiple existing designs. Given that existing designs have already survived the harsh filter of past selection, such large changes are unlikely to produce something successful, but once in a while new combinations might hit the jackpot to a degree not possible with tinkering. Combining radio-wave communication (a telephone), integrated circuits (computer chips), a liquid crystal display (a screen), a speaker, a microphone, and a camera gave us smartphones. Hybridization and fusion are rampant among languages, recipes, or music styles. In the world of biology, the first eukaryotic cells were produced when multiple bacterial species combined into one, just as two eukaryotic species can now hybridize and spark the creation of a new species.

For inheritance, evolutionary biologists Eva Jablonka and Marion Lamb have identified four main modes: genetic, epigenetic, behavioral, and symbolic. Some kinds of variants are inherited in just one way: DNA mutations, for example, are inherited only via genetic transmission. Other kinds of variants can be subject to multiple modes of inheritance: Human behavior, for example, can be influenced by all four. There is no one-to-one mapping between the nature of a new variant (e.g., whether we consider it biological or cultural) and its mode of inheritance.

Genetic inheritance is based on DNA that gets passed from parent to offspring, or from one organism to another via horizontal gene transfer. New variants are introduced by mutation—changes to the nucleotide sequence of a genome. DNA provides the instructions to build a bacterium, a pea plant, a butterfly, or a giraffe. However, DNA is far from the whole story of how biological traits are inherited. In a complex, multicellular organism like a giraffe, all cells have the exact same DNA, but they can have highly divergent shapes, sizes, and functions. When skin cells give rise to more skin cells, or liver cells to more liver cells, the inherited differences are *epi*genetic.

Epigenetic essentially means on top of genetics, and epigenetic inheritance can take many forms. The best-known examples involve new variants created when chemical (methylation) tags get attached to DNA, telling the cellular machinery whether to turn a given gene on or off. These tags are acquired during the development of multicellular organisms, resulting in differentiation of organs, and they can also be triggered by the environmental conditions experienced by whole organisms. Different genes need to be turned on or off to produce a skin cell or a liver cell, and for organisms to survive under certain kinds of stress. Epigenetic tags on DNA can be quite persistent and heritable, if not as persistent as changes to the DNA sequence itself.

At some point in the history of life, animals evolved the ability to learn socially, by observing one another. Birds and whales often learn the songs they sing from hearing other birds and whales, and chimpanzees learn from each other how to break open nuts with rocks. This is behavioral inheritance. A second form of cultural inheritance relies on symbols—usually in the form of language or images. Unlike behavioral inheritance, symbolic inheritance permits the transmission of information about how to do something without needing someone to actually do that thing. For a bird to learn a song, another bird has to sing the song. For a human to learn a song, they have at least two additional options: They can read written music or listen to a recording, in both cases drawing on materials that may have been produced decades or centuries ago. Symbolic inheritance, in the form of language, also goes beyond behavioral inheritance by allowing people to share completely imagined stories and different possible futures, both of which feature prominently in human cultures.

Jablonka and Lamb used this classification of four modes of inheritance largely to help explain the evolution of traits of individual organisms—their shapes, sizes, religious beliefs, or musical preferences. But the different modes also apply at other levels of analysis. Technologies and institutions are not tied to individual people, but the inheritance of their traits still involves behavioral and symbolic transmission of information between people. The divergence of one animal species into two is typically the end point of a long process that might involve behavioral and epigenetic inheritance, in addition to the accumulation of differences via genetic transmission from parents to offspring.

Although variation generation and inheritance are, in principle, conceptually distinct processes, some situations are ambiguous. If the Ford Motor Company adopts a practice copied from Toyota, this could be seen either as inheritance of a trait from Toyota or as the generation of new variation at Ford. Similarly, when one bacterial species incorporates a gene via horizontal transfer from another species, we could consider this to be a form of inheritance (from a nonparent) or a type of genetic mutation. Biologists call such mutations insertions: a new stretch of DNA that gets inserted into the genome, making it larger. In trying to understand evolution in either system, it doesn't much matter what words we use to describe a given situation, but cases like this indicate that some dials on the Evolutionary Soundboard might be interlinked, as we shall soon see.

Despite the many case-specific details of how variation generation and inheritance work, the key inputs to models characterizing any evolutionary system can be summarized with four dials on the Evolutionary Soundboard: two for variation generation, and two for inheritance. These dials allow us to set aside discrete categories of how variation is generated or how inheritance works, focusing instead on how these translate into universally applicable inputs for an evolutionary model. For variation generation, we need to know how often new variants are produced and what their distribution of trait values is. For inheritance, we need to know *which* other entities contribute and *how many* there are: For a given person is it just elders (e.g., parents) or also friends? And how many other people in total?

The Evolutionary Soundboard,
Part I: Variation Generation

Dial 1: The Rate of Variation Generation

The first dial on the Evolutionary Soundboard is for the rate of variation generation (fig. 4.1), which is the key input for all other evolutionary processes. As for most of the dials that will be added to the soundboard, Dial 1 varies on a scale from low to high, minimum to maximum, although the units of measurement will be system specific. Rates of variation generation vary greatly among different evolutionary systems, and they can be tricky to measure.

For cultural entities—tools, technologies, institutions, or languages— one might think that estimating rates of variation generation would be easy, since it is people that are both creating and studying the new vari-

THE EVOLUTIONARY SOUNDBOARD
Part I

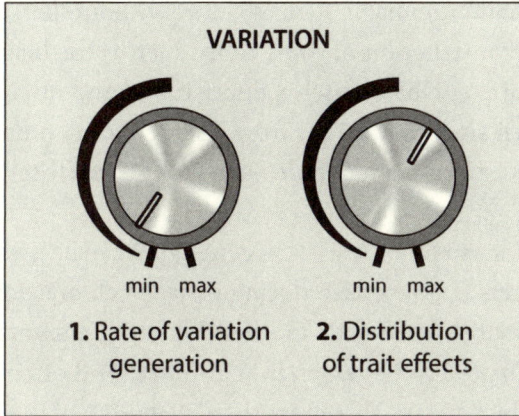

VARIATION

min max   min max

1. Rate of variation
generation

2. Distribution
of trait effects

FIGURE 4.1. Evolutionary Soundboard dials characterizing the process of generating new variants. Dials are set as follows: 1. An arbitrarily low rate of variation generation; 2. Imagining that this dial represents the mean of the distribution of trait effects, here it is biased positive (above the midpoint), as in the lower panel of figure 4.2.

ants. For example, in recent years, the *Oxford English Dictionary* has added more than three thousand new words, senses, and subentries per year, so we can estimate quite precisely the rate of addition. This can only be done, however, at this very specific level of analysis. Words are given serious consideration by the OED only if they are deemed to have "sufficiently sustained and widespread use," which means that they have already been favored by selection to some degree. The number of new words or meanings of words used by anyone anywhere is going to be much higher than the number making it through the OED filter.

Estimating rates of variation generation for most other aspects of culture faces a similar challenge. At certain levels, the accounting is fairly easy, as for products that make it to market or tools common enough to appear in archaeological digs. At other levels, it is much more difficult, as for product designs considered during brainstorming or tool designs only ever tried once or twice. Despite difficulties in making precise measurements, it is clear that the rate at which new cultural variation appears varies widely. As mentioned in chapter 3, Thomas Edison produced more than five hundred sketches of telephones and no doubt imagined many more without sketching them. In contrast, the construction of some tools, such as the blades of certain Japanese swords, involves such a finely tuned and intricate series of steps that even small deviations from protocol can compromise their physical properties. Craftspeople have thus stuck closely to protocol for centuries.

For mutations to DNA, variation generation can be measured at several levels, such as how often, during a given cell division event, one nucleotide is substituted with another. Mutation rates at this level are typically estimated to be less than 0.000001 (and often much lower than that), which means that more than 99.9999% of the time, a given nucleotide in a new copy of DNA will be the same as in the original copy. Considering the size of a genome (three billion base pairs for humans), it is nonetheless quite likely that at least one or two mutations were introduced in the sperm and egg that ultimately joined together to produce you (and me). At the level of an entire population, such as the almost eight billion people on Earth, tens of millions of mutations appear in the 100+ million babies born each year. The DNA mutation rate is highly variable among different parts of a genome, for different kinds of mutation (e.g., single nucleotides versus larger insertions or deletions), and among different species, but in general it is a lot lower than the rate of epigenetic change—for example, the addition or deletion of methylation tags on DNA.

The rates at which new species appear are usually quantified as the average number of new branches on the tree of life that are spawned by a single branch over a given period of time—a million years is used most commonly. By analyzing DNA sequences of many species, researchers can re-create the evolutionary history of different sections of the tree of life, along with estimates of how fast species are produced and how often they go extinct. As one example, in the evolutionary history of fishes, the average speciation rate was estimated to be 0.14, which means that if we start with one hundred species, fourteen new species should be produced over a period of a million years. As with mutation rates, speciation rates vary greatly among different branches of the tree of life, and over time during the evolution of a given group.

Overall, different evolutionary systems demand different units of measurement for the rate of variation generation, but for any general category (e.g., languages or genomes), rates can be highly variable from one instance to the next (e.g., English versus Estonian or bacteria versus mammals). Whether modeling language, technology, or ecosystems, the variation generation dial can be tuned up or down to reflect this, with two main consequences for evolution. First, more variation means more fuel for selection or drift to drive evolutionary change. Second, more variation means more frequent deviations from any optimal design that might have been honed by selection. In other words, adding variation can accelerate adaptive evolution toward an optimum trait value in a population, but if the average trait value is already at the optimum (e.g., the *average* bird beak is just the right size to consume local seeds), then adding variation results in more *individuals* having suboptimal trait values (e.g., beaks too small or too large for the local seeds). This means that the rate of variation generation is itself a trait that can influence long-term evolutionary success and thus be under selection. Too little variation, and there is no scope for adaptation; too much, and average fitness can be dragged down.

Dial 2: The Distribution of Effects

Some things that capture our attention most intensely when observing the workings of the world are not actually so important when we sit down to build a model of how that world works. As discussed earlier in the chapter, many observers have been struck by the seemingly glaring distinction between randomly produced variation in biology and deliberately produced variation in culture. So, we could imagine adding a dial to the Evolutionary Soundboard specifying where a given evolutionary system falls on the continuum between the two extremes. We might also imagine a dial that toggles between the tinkering and combining modes of variation generation. But if we want to predict the course of evolution— how traits will change over time and how fast—these distinctions matter only to the extent that they alter how much new trait values deviate from old ones. We need to know whether new word variants tend to be shorter or longer than the old ones, but not whether those variants arose by random spelling errors or by deliberate efforts from the language police. What we need is a dial that characterizes the distribution of trait values among new variants (Dial 2 in fig. 4.1), regardless of how those variants were produced.

The specific traits of interest will be highly system specific. For a hand axe, the traits might be durability, sharpness, or size. For a bacterium, they might be the degree of antibiotic resistance, metabolic rate, heat tolerance, or cell size. When variation is produced by random accidents—be they errors in tool construction or DNA mutations—we might expect a distribution of effects centered on zero. That is, a new axe or bacterial cell is just as likely to be a bit bigger as it is to be a bit smaller than the earlier ones.

If we assume that the original trait value is reasonably well adapted to its circumstances, then we might also expect that any new variants hanging around long enough for us to measure them will not be too different from the original. So, we might also expect small deviations from the baseline to be found more often than large deviations. We don't expect to find a 10-meter-tall axe in an archaeological dig or a balloon-sized bacterial cell floating around the ocean. So, the distribution of trait effects might look like the top panel of figure 4.2. Variation produced by tinkering involves small changes and so would make for a relatively narrow distribution. Variation produced by combining should make it broader—more often producing trait values far from the original.

For intentionally produced variation, the distribution of effects is not likely to be centered on zero, because in this case, traits are systematically pushed in one direction or the other. People might deliberately produce hand axes that are more durable (moving the distribution in fig. 4.2 to the right, if the x-axis is durability) but also smaller (moving to the left if the x-axis is size). Scientists might genetically engineer bacteria to be more heat tolerant, but with a lower metabolic rate. Bacteria incorporating viral DNA into their genome might become system-

Random generation of new variants

Biased generation of new variants

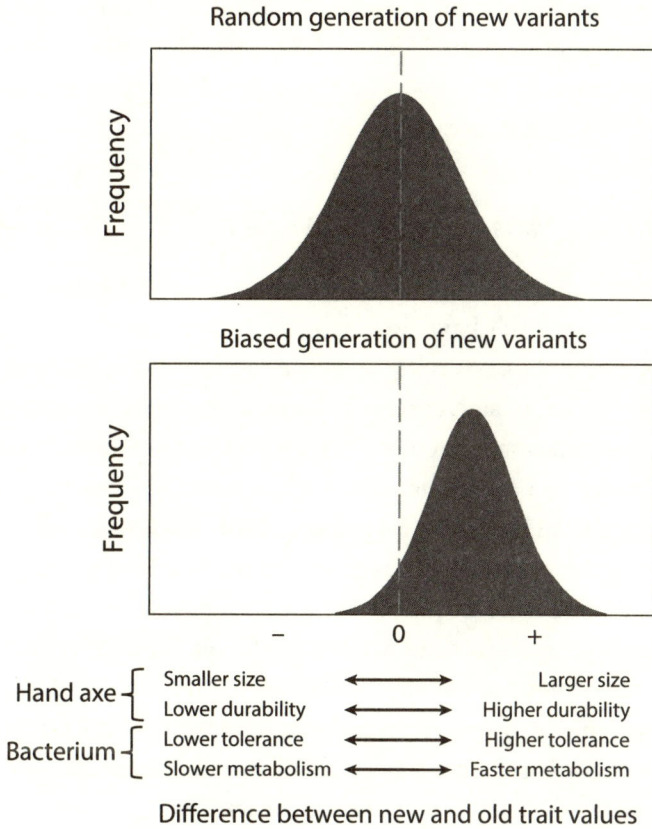

FIGURE 4.2. Hypothetical distributions of trait differences between newly generated variants and the original values. The bottom panel might represent people deliberately introducing variants of a hand axe that are larger or more durable than the average existing axe, or a tendency for mutant bacteria to have greater heat tolerance or metabolic rate.

atically more resistant to infection, rather than less resistant. Although directed variation might cause a distribution of trait differences that is biased toward larger values (bottom of fig. 4.2) or smaller values, observing such a distribution does not necessarily mean there was any deliberate process involved. Even randomly created variation might produce differences that aren't centered on zero, based on physical constraints. For example, if the size of a stone axe is limited by the size of available rocks as starting material, random construction accidents might be more likely to produce smaller axes than larger ones. Random mutations might be more likely to reduce the efficiency of an enzyme than to increase it. The important things are the shape and position of the distribution, not its cause.

Are New Variants Better or Worse than the Old Ones?

Given that the traits of interest are often unique to a particular study system (plants don't have fins; fish don't have leaves), evolutionary biologists often bypass traits altogether and instead study the fitness consequences of whatever new mutations arise in a population. This provides a common currency with which to compare such distributions for any and all species, even when those species have little in common in terms of traits (it applies to both fins and leaves). And we can take the idea further, as it is generally applicable to all evolutionary systems, as long as we can assess the success (fitness) of new variants of tools, cell phones, or recipes, relative to the old. The concept of evolutionary fitness has been briefly mentioned in previous chapters and will feature prominently in chapter 5 (on selection). Given its importance for understanding the consequences of variation generation, we need to take a brief dive into the concept of fitness here.

While on one hand fitness is an intuitive and straightforward concept, it can also be quite slippery and elusive. In its most general sense, fitness is the degree of success expected for a given type of entity. But as we saw in chapter 3, there are multiple ways of defining success, based partly on different definitions of what constitutes an entity: For a company, success can be measured as market share or profits; for a slime mold, as areal coverage or DNA copies. There is also the open-ended question of the relevant time frame: A company, product, genotype, or species might enjoy great success over the next year (or day, or month, or decade, or generation), but failure in the one after that. Limitations on how frequently we can make observations will determine whether we see fitness going up and then down or if we see it remaining constant. This elusiveness pervades biology itself—that is, it is not a consequence of generalizing evolutionary theory. Philosophers and scientists have

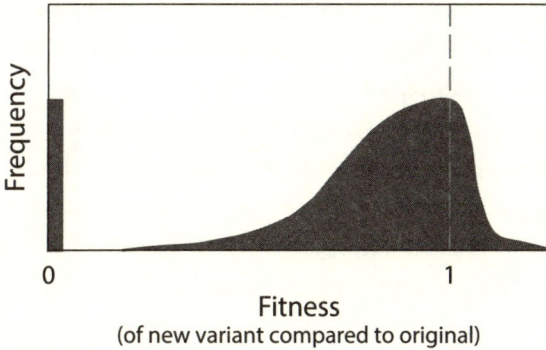

FIGURE 4.3. A stylized distribution of fitness effects of new genetic mutants. A value of 0 indicates inviability (the new organism cannot survive); a value of 1 indicates no fitness difference between the new mutant and the original type.

tied themselves in knots trying to pin down a definition of fitness, but for our purposes, the general sense just described does the job. Here we can think of the fitness of a new variant as its expected degree of success over whatever unit of time works best for the situation.

In general, altering the design of something that is itself a product of evolution—i.e., that succeeded where many others failed—is more likely to do harm than good. In short, most new stuff sucks, whether it is a new music band or a mutant fruit fly. For short-lived beasts in the lab like bacteria, fruit flies, or viruses, biologists can generate thousands of mutants with small DNA alterations and then assess their ability to survive and reproduce, relative to the initial type. Inability to survive means fitness = 0. If the new fitness is the same as the old, we say that relative fitness = 1 (new fitness divided by old fitness). The fitness distribution of random DNA mutants often looks something like figure 4.3.

The two most common fitness outcomes of a new random mutation are death or no change at all. For relatively small genomes in which most parts have some function—e.g., in viruses—many mutations disrupt function to the point of inviability: fitness = 0. At the same time, the genetic code that translates DNA sequences into proteins has a great deal of redundancy, meaning that many nucleotide changes have no functional consequences in terms of the protein produced. For more complex organisms like pea plants or people, much of the genome codes for nothing, so an even larger portion of mutations causes no change in fitness. When a mutation has some nonlethal effect, often it is mildly negative, occasionally positive. The result of all this is a distribution with many inviable new variants, many with roughly the same fitness as the original type, and otherwise more new variants with lower fitness than with higher fitness (fig. 4.3).

For other kinds of variation generation, we don't have precise quantifications of the distribution of fitness effects, but we have good reason to expect something similar to figure 4.3, at least qualitatively. In the realm of technology, for example, if the current design works well, most small changes one can imagine will not represent an improvement, and some will render the technology—an engine, a hydroelectric dam, or a computer—entirely nonfunctional. A few changes will improve the function (and/or the desirability to consumers), and fewer still will represent a major improvement. Research and development teams are charged with finding those needles in the haystack.

When new variants are directed rather than random, the hope of the "directors" is to reduce the number of new variants that are harmful and to increase the number that are helpful—that is, to shift the fitness distribution (fig. 4.3) to the right. Whether it's a bacterium taking on viral DNA, an environmental trigger causing gene regulation via methylation, a genetic engineer attempting to add an herbicide-resistance gene into a plant, or a product designer putting forward ideas, variants are being generated for which there is reason to anticipate positive fitness consequences. Of course, all these kinds of direction are highly imperfect, with no guarantee of positive results. Bacteria can end up attacking their own DNA, gene regulation might lag behind environmental change, genetic engineers create a great many inviable plants, and many product launches fail miserably. Directed generation of variation might tip the odds a bit in favor of something good, but it might not do much more than that.

To summarize the gist of Dial 2 on the Evolutionary Soundboard, variation generation matters to the evolutionary process to the extent that new variants have new and different traits. It is the shapes of the distributions describing trait and fitness differences that matter, not their underlying causes, interesting as those causes might be. Dial 2 determines the distribution of trait effects (fig. 4.2). The distribution of fitness consequences of new variants (fig. 4.3) depends additionally on a dial that will be introduced in the next chapter—one that determines the relationship between traits and fitness. In mathematical terms, we require more than a single number to summarize distributions like the ones in figure 4.2, but here we can just think of Dial 2 as standing in for a bit more complexity. The complexity is needed only when writing down actual equations—a task we can leave to the mathematicians.[2]

The Evolutionary Soundboard, Part II: Inheritance

My own genome was inherited from two people: my mom and my dad. In contrast, my cultural practices and beliefs—and especially my knowledge—have been inherited, and continue to be inherited, from a much broader set of people, some of whom I do not know personally. The genome in a bacterial cell was also mostly inherited from its "mother" cell after it split in two, but parts of the genome might have been acquired from other cells—possibly of different species—via horizontal gene transfer. Although only a small fraction of a given bacterial cell's genome will have been acquired by horizontal gene transfer since the most recent cell division, on average 80% or more of the genes in its genome have been involved in horizontal gene transfer at some point in the past. General Motors circa 2024 will have inherited most of its business practices and routines from earlier iterations of the company, with a healthy dose of lessons gleaned from other successful companies as well. For all these scenarios of inheritance, two key variables can capture their similarities and differences: the "direction" and the number of influences.

Dials 3 and 4: The Direction of Inheritance and the Number of Influences

Dial 3 on the Evolutionary Soundboard characterizes the degree to which inheritance is entirely vertical—that is, from antecedents (e.g., parents to offspring)—versus also including horizontal influences from contemporaries (fig. 4.4). Antecedents are older or previous versions of an entity, be it a musical style, an elephant, or a business firm. For almost all models in biology, this dial is tuned up to 100% vertical and therefore treated just as a background assumption rather than a formal part of the model. The Second Science brings this dial to the surface. We now know that many biological systems require this dial to be tuned down, and cultural systems vary widely in the degree to which inheritance is dominated by either the horizontal pathways (e.g., fashion choices) or vertical pathways (e.g., religion).

The second dial for inheritance, Dial 4, characterizes the number of entities from which one focal entity has inherited its characteristics. An inventor might design a new tool using just a single previous tool as a template, or they might take lessons from a large range of other tools. In the absence of horizontal gene transfer, the genome of a bacterial cell is inherited from just one other cell, its "mother." Genetic inheritance in sexual eukaryotes involves two parents, while adding horizontal gene transfer in either case increases that number beyond one or two.

On the surface, it might seem like horizontal inheritance necessarily increases the number of influences, but the two dials are independent. A child can learn cultural norms from a large number of elders (entirely vertically) and might develop a new musical preference from just one peer (horizontally).[3] We can also imagine two individual plants that both inherited their genes from two other organisms (Dial 4 = 2), but with different vertical versus horizontal contributions. The first plant, produced via sexual reproduction, inherited 100% of its genes vertically from its two parents (Dial 3 = maximum), while a second plant, produced clonally, might have inherited genes both from its one parent (vertically) and from a different species horizontally (Dial 3 < maximum).

Together, these two dials capture the key features of how traits are inherited by entities in any and all evolutionary systems. Tuning down Dial 3 (increasing the importance of horizontal influences) or tuning up Dial 4 (increasing the number of influences) should both have the consequence of accelerating the evolutionary process. Vertical inheri-

THE EVOLUTIONARY SOUNDBOARD
Part II

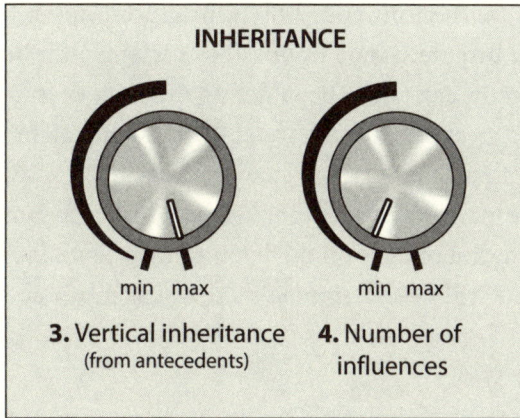

INHERITANCE

min max   min max

3. Vertical inheritance   **4.** Number of
(from antecedents)   influences

FIGURE 4.4. Evolutionary Soundboard dials characterizing inheritance. Dials are set to their usual assumed values in biological systems: 3. Only vertical inheritance, and 4. Few influences (one for clonal reproduction or two for sexual reproduction).

tance alone permits relatively slow evolution, requiring the passage of generations before we will see substantial changes in the prevalence of different variants. Horizontal inheritance can happen much faster. Likewise, if the number of influences on any one entity is increased, there is necessarily also an increase in the number of entities that any one can potentially influence. I have only two parents and two children of my own, but I have a great many scientific influences, some of whom have influenced millions of others. This means that my scientific ideas can potentially spread more quickly than my genes.

Taking Variation and Inheritance Seriously

People love a neat and tidy story. But people can't always get what they want. By treating variation generation as entirely random and inheritance as entirely vertical, the modern synthesis of evolutionary biology was able to simplify models greatly, leaving one main creative force in evolution: natural selection. A more generalized theory, covering the

full range of evolutionary systems—both outside and inside biology—
requires paying more attention to how the distribution of traits among
new variants can be influenced by nonrandom direction and other
factors, and a broader range of possible systems of inheritance. The
underlying details can take a bewildering diversity of forms, and scientists still have a great deal to learn about their inner workings—especially
in the realm of epigenetic, social, and cultural systems. But the consequences of the many ways variation is generated can be summarized suc-
cinctly in a handful of dials on the Evolutionary Soundboard. The *fates*
of new variants will be determined by the dials introduced in the next
two chapters.

5

Selection, "Natural" and Otherwise

At various times over the past eleven to twelve thousand years, human populations in several parts of the world made a transition from a hunter-gatherer lifestyle to a more sedentary agricultural lifestyle. It is difficult to know how exactly these transitions happened, but archaeologists, anthropologists, and biologists have pieced together the broad outlines of a likely scenario. One thing is clear: It would have been nearly impossible for one or a few visionaries to go from their knowledge of wild food sources to imagining a future in which wild plants and animals had been transformed into divergent forms and grown at high density in large human settlements. Rather, both the cultural and biological aspects of agriculture evolved gradually, during centuries or millennia of trial and error.

Domestication of food plants probably started with people collecting seeds and fruits from wild plants and bringing them back to their settlements, temporary as those settlements might have been. The simple act of collecting involves selection (in the evolutionary sense): Seeds and fruits that are bigger and tastier are more likely to be collected, as are the ones that stay on a plant for longer when ripe. Not everything brought home gets eaten, and much of what does get eaten still results in seeds being deposited nearby after passing through our guts. All this would result in more of people's preferred types of plants sprouting up near their settlements. Those plants near at hand would, in turn, be part of the set from which people next selected the most desirable ones—whether blue-berries, pea pods, or grass seeds. Slowly but surely, people would have started tending to plants, noticing rare ones with desirable properties (e.g., an almond without the bitterness of most wild ones), planting seeds from the best of the last batch, and so on. Things eventually got quite deliberate, but much of the early selection would have been unintentional. As explained by Jared Diamond, "Food production *evolved* as a by-product of decisions made without awareness of their consequences."

Human agriculture has also involved profound cultural evolutionary changes. Growing food in surplus quantities in small areas selected for denser and more permanent human settlements, as well as for spe-cialization of trades, given that not everyone needed to be involved in food production. Strong selection for an increased and more reliable food supply sparked the evolution of technologies for ploughing, seed-ing, irrigation, and surplus food storage. Many rounds of scientific trial and error—variation and selection—culminated in scientists of the early twentieth century figuring out how to convert nitrogen from the air, which is highly abundant, into a solid form that can be used as fertil-izer. Enhanced fertilizer supply helped spark the Green Revolution, which produced massively increased yields of novel crop varieties in many parts of the globe. The cultural evolution of agriculture has, in turn, fed back to influence human biological evolution. Altered diets and food preparation practices (more cooking) prompted the evolution of weaker jaws in modern humans compared to their hunter-gatherer ancestors (cooked food is softer than raw food). Where dairy farming became prevalent, lactose-laden milk selected for human genes that conferred the ability to digest lactose as adults (in most of the human population, the ability to digest lactose declines with age).

The story of agriculture involves many cultural and biological sub-systems, within which scientists have applied a variety of labels to de-scribe the action of selection: natural, artificial, genetic, behavioral, cultural, individual, group, and multilevel, among others. This chapter is about selection, and one of the key points is that a broad conception of selection—which we will use in the Second Science—renders most of these finer distinctions unnecessary. Take the distinction between artificial and natural selection. For biologists, selection is considered "artificial" when the selecting is done by people, who consciously and intentionally decide which plants or animals will contribute to the next generation. Otherwise, it is considered "natural." This is all fine and good, but the distinction is, well, somewhat artificial. Domestication seems like the most clear-cut case of artificial selection—done by people, done with intent. However, much of the selection done by people is not at all intentional (as we have already seen), and in fact, humans are not the only animals to have domesticated agricultural crops.

In many neotropical forests, millions of leaf-cutter ants move constantly in bidirectional traffic along miniature manicured roads. The ants headed away from home are going toward a source of leaves, while those headed back are carrying sections of leaf blade several times their own height. When they get home, the ants will deposit their leaves somewhere in a network of thousands of connected chambers, where the leaves feed colonies of fungi. The resulting fungal crop will, in turn, feed the ant larvae. Special swellings on the fungal hyphae called gongylidia are the most prized portion of the fungus to use as food. More than two hundred species of ants take part in ant-fungus symbioses of this type, which originated some sixty million years ago. Humans were not the first agriculturalists. Not by a long shot.

Domestication happens when cultivation by one species (e.g., an ant) exerts selection on another (e.g., a fungus), causing genetic change that is of benefit to the cultivator. In the most sophisticated of the ant-fungus agriculture systems, the fungi show clear signs of having been domesticated by ants. Not only is the fungal partner now entirely dependent on the ants for survival, but compared to noncultivated fungi, their production of enzymes is better suited to breaking down fresh leaves, and they more reliably produce gongylidia. In short, the agricultural practices of ants have been a dominant force of selection on the fungi they cultivate, and this is just one example of animals imposing selection on other species. Every sweet and fleshy wild fruit, like raspberries, cherries, or mangos, evolved under selection imposed by animals making decisions about which fruits to eat, and so which seeds got dispersed to suitable places for germination. All these animals' decisions were in a sense intentional, but we still call it "natural" selection because the decisions were not made by people.

The distinction between natural and artificial selection gets even fuzzier when we consider that selection imposed inadvertently by human actions is sometimes called natural (e.g., when the action is urbanization) and sometimes artificial (e.g., when it is trophy hunting). Trophy hunting might be considered artificial selection because people make decisions consciously about which animals live or die, but when human hunters kill bighorn sheep with the largest horns, they do not *intend* to influence the evolution of horn size any more than wolves intend to influence the evolution of running speed by catching the slowest prey. The question of whether ants or birds or lions are conscious of their decisions and their consequences in the way that humans are might be interesting to a psychologist, but it is of no special relevance to understanding evolutionary dynamics. Similarly, knowing that humans are an agent of selection might be inherently interesting to other humans, but of no more significance to understanding evolution than knowing that the agent of selection is elephants, urbanization, or long winters. As was the case for random versus directed generation of variation, not all distinctions that capture our attention (in this case, natural versus artificial selection) are necessarily the most important when trying to understand how evolution works.

In the rest of this chapter, we will first define selection in a maximally general way, so that it can apply to all types and levels of selection recognized by evolutionary scientists. This will involve revisiting (from chapter 3) the importance of recognizing the many levels at which evolutionary dynamics can occur. We will then outline different forms of selection that apply across the full range of biological and cultural scenarios. These forms of selection will be incorporated into several new dials that will be added to the Evolutionary Soundboard.

Selection: Systematic Differential Success

In the recipe for evolution provided in chapter 3, the third ingredient is "entities have differential success." One component of this differential success is random and will be treated in chapter 6. The other component is systematic, or predictable based on an entity's traits: We call it *selection*. We can thus define selection as *systematic differential success among entities based on their traits*. This paraphrases a beautifully concise and generalized definition of selection provided by philosopher and historian of science Eugene Earnshaw-Whyte: "the systemic advantage of a type." We have replaced *systemic* with *systematic* (a more precise match to the concept), and *advantage* with *differential success* (an advantage for one type implies a disadvantage for others). And we can think of *types* as being defined by any *trait(s)* of any kind of entity: a transposon with a given sequence, a gene with a given phenotypic effect, a cultural preference, a tool design, a business practice, or a military strategy.

Earnshaw-Whyte also provides a generalized definition of evolution itself—"the trajectory of trait-prevalence over time." This generalized conception of evolution requires that we expand the idea of selection to include certain things that biologists would not usually associate with natural selection. For example, success need not involve reproduction, and if evolution is happening at multiple levels simultaneously, selection can cause an increase in the prevalence of a given trait even if the individual organisms (or businesses, or cultural groups) that possess the trait have reduced survival and reproduction. On the surface this is counterintuitive. Some examples can help illustrate how observations from nature and culture necessitate a conception of selection that is broader than the typical version in biology.

In your body, two nonidentical copies of each chromosome are present in every living cell, except for the gametes—sperm or eggs—which contain just one copy.[1] When sperm and egg cells are formed, genes are first swapped between the two parental copies of a given chromosome in a process called recombination, resulting in a unique DNA sequence in each gamete. Most of the time, each allele in a gamete's genome is equally likely to have come from either one of the parental genomes. But not always. Many examples have been found in which one of the two parental alleles can tip the odds heavily in its own favor.

In mice, the "t haplotype" is one version of a large region of chromosome 17 within which recombination is strongly suppressed, such that the whole region tends to be inherited in one piece. If a male mouse has one copy of chromosome 17 with the t haplotype (allele) and one copy with the normal sequence, sperm with the normal sequence will have severely reduced function. As a result, almost all successful fertilization events involve the t haplotype. In essence, the t haplotype strongly suppresses and so outcompetes the normal type at the stage of sperm development; this is clearly *systematic differential success*. For a whole mouse (not just a sperm cell), however, there is no advantage to having the t haplotype. In fact, there is a heavy cost: Male mice with two copies are sterile. So, at the level of genotypes of mice, there is, on average, a *dis*advantage of having the t haplotype. Although it seems clear enough that selection favors the t haplotype during sperm development, biologists do not call this selection: They call it *gene drive* (of which there are many other examples). The influence of the t haplotype on mouse fertility is, on the other hand, described as an example of natural selection. In the Second Science, both are clear examples of selection, just operating at different levels and at different phases of a mouse life cycle.

Another way to describe gene drive is to call it *transmission bias*: altering the odds that a particular type gets transmitted from one generation to the next. Transmission bias is common in cultural evolution. As you and I grew up, the cultural traits we inherited, such as our political views or our work habits, were not influenced by a random sample of the people we knew. In many cases, we are more likely to copy or emulate elders, people of high prestige, or people who are more like us in other ways (e.g., how they look), even if different choices might better suit our needs or desires (e.g., career success). In an experimental setting, children learning how to use a toy by watching others were much more likely to copy a person who was given the appearance of having "prestige," in the form of adult bystanders paying them more attention. More generally, people in power can have a massive influence on the transmission of cultural traits. The fact that such transmission biases can conflict with some of our needs or desires has prompted cultural evolutionists to distinguish transmission bias (nonrandomness in who gets emulated) from selection (nonrandom behavioral choices based on meeting certain needs). However, both situations involve selection in the broad sense, just at different levels or at different phases of an evolutionary process.

At the same time that evolutionary scientists have excluded some selective processes, such as transmission bias, from the concept of selection, they have also excluded some kinds of selection from being considered "natural." In addition to drawing a distinction between natural selection and artificial selection, biologists also distinguish at least two other kinds of selection: sexual selection and group selection.

Sexual selection involves mate choice. For centuries, naturalists have been intrigued by the elaborate, colorful structures on some animals—birds in particular—that are far more developed in one of the two sexes, most often males. The peacock's tail is the classic example, but striking (if less ostentatious) examples abound: Outside my backyard window, the bright red cardinals and brilliant yellow goldfinches are all males. The females of both species pale in comparison (pun intended). Bright colors ought to make a bird more visible to predators, and a peacock's tail surely can only be a hindrance to moving about, gathering food, or avoiding getting eaten. In other words, these traits should not be favored by *natural* selection, and yet the flashiest peacocks nonetheless have the most offspring.

The peacock paradox can be solved by invoking *sexual* selection: preferences of females for flamboyant males as mates. Regardless of which sex is doing the selecting, it is impossible not to also see sexual selection in action on human traits under cultural inheritance, such as fashion choices and other practices and norms related to physical appearance. Studying sexual selection requires some special considerations, such as the need to account simultaneously for the evolution of male appearance and female preferences. But as with artificial selection, a category was created to isolate a particular agent of selection (mate choice), of which there might be many others, such as conflicts between parents and their offspring in terms of how much energy to expend on parental care. Sexual selection is just one of a great many special cases of selection.

In the cases of natural, artificial, and sexual selection, there is typically an assumption that selection is based on differential success among individual organisms and their genes. The question of whether selection can also happen among *groups* of entities, driven by the differential success of those groups (e.g., populations of animals), has been subject to debate. Consider colonies of ants, where most of the workers are actually sterile, with no chance of reproducing. If natural selection favors traits that confer success to individuals, how can we explain the evolution of so many sterile workers? The conundrum was solved by realizing that the workers are actually promoting the success of their *genes*. Due to a quirk of genetics, ants in a colony share more of their genes with one another, on average, than do human siblings. So, if the colony overall is successful, sterile workers are, in a sense, passing on their genes to future generations, even if just a few of their brothers and sisters do the actual job of reproducing. If a gene prompting workers to protect and feed the egg-laying queen promotes her genetic success, that gene will be favored by selection.

More generally, the term *kin selection* describes the process by which seemingly selfless behaviors are favored in individuals because they promote the success of relatives (kin), with whom those individuals share more genes than they do with unrelated competitors. Some scientists thus view these individuals, and their genes, as "selfish," so that even what seems like an obvious example of selection favoring cooperative *groups* actually reduces to selection at the level of selfish gene copies.

Reducing an ant colony to a bag of selfish genes is a bit of a sleight of hand, as it remains the case that individuals and their genes succeed only if the group succeeds. The division of labor (including who gets to reproduce) is a characteristic of the colony, not of any one individual, and the evolution of the group characteristics hinges on the relative success or failure of different colonies. And the closer one looks at nature and culture, the more one sees evolution happening at multiple hierarchical levels. Eukaryotic cells evolved when coalitions of multiple prokaryotic cells outperformed their component parts. Multicellular eukaryotes (like humans) evolved when coordinated groups of cells outperformed individual cells. Cultural traits can spread not only via the influence of individuals on one another within a given society, but also because some whole societies displace others. Fully understanding many systems (e.g., human culture) requires recognition that selection happens at multiple hierarchical levels.

In the Second Science, a multilevel evolutionary system is approached by first considering each level separately and then drawing links across levels. For each level—let's say individual people within a society, and multiple societies—we can tune the dials on the Evolutionary Soundboard to the appropriate positions, and we can allow some dial positions to depend on what's happening at the other level. The expected success of a society, for example, will depend in part on characteristics of people as individuals. Often the two levels are found to conflict. Selection within a society might favor more selfish individuals with characteristics that allow them to secure the most resources for themselves. In contrast, selection at a higher level might favor societies with greater degrees of cooperation. Evolutionary biologist Mark Pagel refers to tribal groups as "cultural survival vehicles," within which cooperation can help people to survive in harsh conditions (regardless of any competing groups), to succeed in warfare, to have their practices copied by other groups, and to attract other people to join their group. Many religious or cultural rituals are difficult to explain, except as benefiting the cultural fitness of groups.

Regardless of the qualifiers we attach to the concept of selection—natural, artificial, sexual, kin, genetic, individual, group, behavioral, or cultural—there are just two main forms of selection that determine the expected trajectories of trait change and diversity. Some new dials on the Evolutionary Soundboard are needed to characterize these forms of selection.

The Evolutionary Soundboard, Part III:
Differential Success—Selection

Compared to a small shrub, a tall tree typically has high fitness in warm and wet places and low fitness in the Arctic tundra or the Sahara Desert. Winter clothing retailers have higher fitness in Canada than they do in Panama. In short, selection means that fitness—the expected degree of success—is a function of an entity's traits and how well those traits match the environment. To simplify matters, we can start by ignoring variation in the environment and consider selection operating at one place and time—a tropical forest *or* a desert, Canada *or* Panama—in which there is just one "environment." Given a single environment, fitness then depends only on traits.

Dial 5: The Fitness Function

All evolutionary models require a function that relates trait values to fitness. The relationship can be simple or quite complicated.

In 2007 Apple released the first iPhone, and within a few years iPhones were one of the dominant products in the cell phone market, helping trigger steep declines in sales of some other brands, such as BlackBerry and Nokia. The rapid rise of the iPhone was likely based on selection for a constellation of traits, including the touchscreen, on-screen keyboard, and ease of internet access. At the time, the fitness function would have looked something like figure 5.1. The same fitness function would have applied when Google displaced virtually all other search engines in the early 2000s. Similarly, in late 2021, when I was pondering whether to write this book, a new variant of the virus causing Covid-19 arose—the Omicron variant. The rapid spread of Omicron in Canada (and elsewhere) was an indication of its higher fitness than the previously dominant Delta variant, and the same fitness function would have applied here as well, based on Omicron's higher rate of person-to-person transmission.

If we extend these simplified examples with just two types to situations with a broad range of types, we might also see a similar straight-line fitness function. We call cases like this *directional selection* because selection always pushes in the same direction (the gray arrows). It might apply to viruses that vary in their ease of transmission, phones that vary in their screen quality or speed, or variants of common words of different length, among many others. The biomedical engineer Hadi Nia and colleagues have shown that the shape of the sound hole on violins likely evolved over a period of centuries under directional selection for air resonance power efficiency (basically, sound volume and quality; see fig. 5.2). Violin makers did not know exactly why some hole shapes were

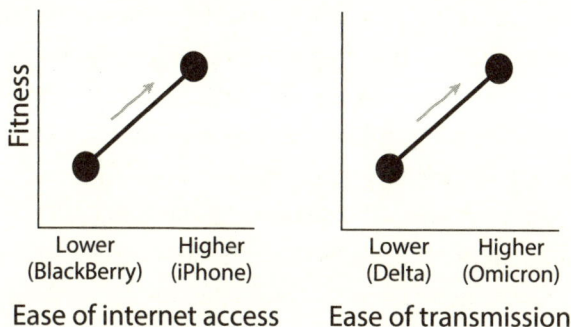

FIGURE 5.1. Directional selection favoring one variant over another based on their traits: ease of internet access for cell phone brands, or ease of transmission of Covid-19 variants. The arrows indicate the expected direction of change in the trait average in each population.

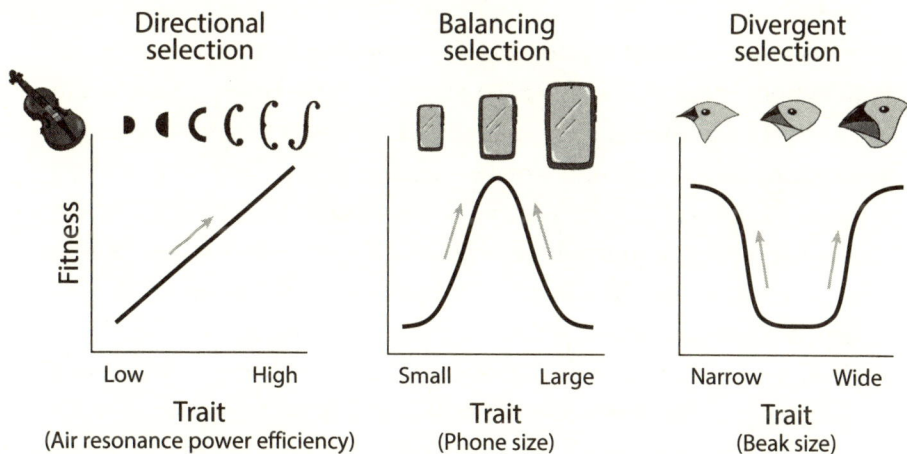

FIGURE 5.2. Possible shapes of fitness functions, with examples: sound hole shape on violins (directional selection), cell phone size (balancing selection), and finch beak sizes (divergent selection). Image of hole shapes in the left graph used with permission of The Royal Society (UK), from "The evolution of air resonance power efficiency in the violin and its ancestors," H. T. Nia et al., Proc. R. Soc. A. 471:20140905, 2015; permission conveyed through Copyright Clearance Center, Inc.

better, just that some happenstance variants worked better than others. The holes on the better-sounding violins were retained.

Directional selection—along with variation generation—explains how sophisticated adaptations can evolve. Violins, viruses, and cell phones are each exquisitely adapted to specific functions, and all of them evolved via innumerable incremental steps. At each step, new variants arose—slight modifications to the hole shape, a new combination of components, a change in size—and selection favored some and not others. Each new version bore a strong resemblance to its immediate predecessor, but over many rounds of variation generation and selection, design changes were profound. Variation generation ensures that the system doesn't get stuck at one trait value, and selection moves the system to new average trait values. The word "adaptation" refers both to the products of this process—the violin, the virus, the phone— and to the process that produced them.

It's also important to know that selection doesn't always push in one direction. When the first mobile phones were produced, they would not have fit comfortably in any pocket that wasn't bizarrely large, and so early selection favored smaller models. But nobody wants a cell phone the size of a fingernail, and so the fitness of a cell phone model is greatest at a size somewhere between a fingernail and a loaf of bread. In a case like this, the direction of selection at any moment in time depends on the range of traits present in the population. When most phones were large, selection favored the relatively small ones. When the average phone got to be smaller than your hand, there was probably selection for larger phones, leading to the in-between size of the average phone we see circa 2025 (as big as or a bit bigger than your hand). There were certainly factors other than size involved (such as screen functionality), and there is almost certainly not one precise optimum size for all phones (people have different preferences), but this example serves to illustrate a very common occurrence: *balancing selection*. If a body part is too big or too small, if a vehicle is too light or too heavy, or if a company invests too much or too little in employee benefits, success is often less than it is somewhere in between. Selection can balance out the extremes.

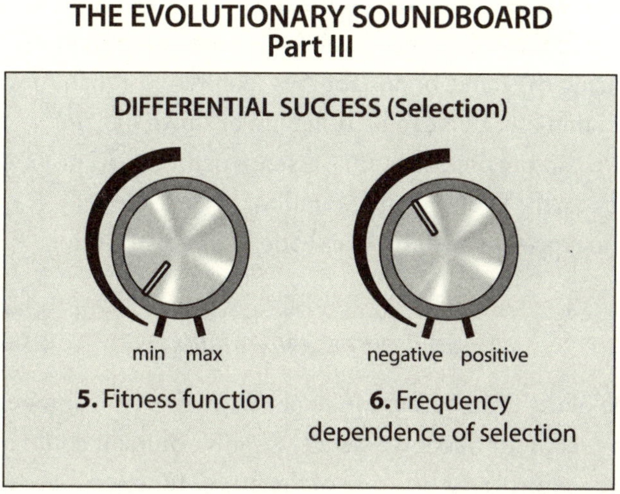

THE EVOLUTIONARY SOUNDBOARD
Part III

DIFFERENTIAL SUCCESS (Selection)

min   max   negative   positive

5. Fitness function   **6.** Frequency
dependence of selection

FIGURE 5.3. Evolutionary Soundboard dials characterizing selection, which is one pathway to differential success of variants. Dials are set arbitrarily: 5: Imagining that the dial represents the slope of directional selection, here it is weak selection; 6: Weak negative frequency-dependent selection (assuming a vertical setting is zero).

In the cases of directional and balancing selection, there is only one fitness "peak"—either at the edge or in the middle of the range of possibilities. In other cases, fitness functions can have multiple peaks, in which case we have *disruptive selection*. If the seeds available to birds are either large or small but rarely anything in between, we might expect selection to favor seed-eating birds that have small or large beaks, but not those with medium-sized beaks. In the current market for written fiction, one can find many collections of short stories and many full-length novels but relatively few fictional works of intermediate length (novellas), possibly due to selection based on reader preferences or economics (e.g., the reduction in production cost for a novella might be less than the reduction in the price people are willing to pay).

Although these three shapes to the fitness function—straight line, one peak, two peaks—make different predictions about the direction of evolution (the gray arrows in fig. 5.2), all shapes can be produced by tuning the parameters of a mathematical equation relating fitness to the different possible trait values. The fitness function can take any form at all. As was the case for the distribution of trait values for new mutations (Dial 2), characterizing the range of possibilities requires an equation with more than one parameter that can be tuned up or down like our soundboard dials. But to capture the conceptual essence of the situation, we can imagine one dial on the Evolutionary Soundboard (left side of fig. 5.3) that can be tuned to represent any possible shape of this distribution.

Environmental Variation

The shape of the fitness function depends on the environment. In fact, in the context of the Second Science, the environment could be *defined* as whatever controls the position of the fitness function dial on the Evolutionary Soundboard. Hot and dry places favor lizards over amphibians; cool and wet places, the opposite. Economic downturns favor discount retail stores, while economic booms favor higher-end retailers.

If selection pushed in the same direction everywhere and always, the world would be a pretty boring place: similar languages, cultures, trees and birds everywhere. But the Earth is far from boring. Environmental variation from one place to another creates different fitness functions in different places, favoring one trait here and a very different one there, and so is a major contributor to the diversity of life and all its products. We've already discussed some clear examples of environmental variation, such as the climate in Canada versus Panama, but selection can also be altered by more subtle environmental variation. Quirks of geology or pollution from mining activities lead to some soils containing high concentrations of heavy metals, selecting for very different kinds of plants— both within and among species—than on "normal" soils only a few meters away. Socioeconomic conditions in nearby city neighborhoods can select for divergent cultural practices, such as the degree of antisocial behavior, although cause and effect are difficult to assess in these situations. Over the long term, these opposing directions of selection in different places can create diversity, by prompting a split of one species or one cultural practice (e.g., religion) into two. In the next chapter we will take up the question of how movement between places influences evolution. For now, we can take the lesson that environmental conditions—and so the shape of fitness functions—can vary tremendously from one place to the next, often acting as an engine for the creation and maintenance of diversity.

Environmental conditions can also vary over time in a single place, continually turning the fitness-function dial clockwise or counterclockwise. A relatively hot and dry summer might be good for snakes, grape growers, and outdoor restaurants, while a cooler and wetter summer is better for frogs, cranberry growers, and indoor restaurants. For any given type of animal, farmer, or restaurant in an evolutionary system there might be good times and bad, and the sequence will not be the same for different types. Much like spatial environmental variation, temporal environmental variation can help to maintain diversity—giving a chance to different types at different times—but the temporal effect is less powerful. If a snake or a grape grower or a restaurant owner suffers severely in cold and wet *places*, they can just avoid those places. But once established in a given place, the cold and wet *times* (e.g., years) can't be avoided and might spell doom, thereby reducing the diversity of types.

Thus far, our discussion has mostly assumed that the environment is something imposed on evolutionary systems from the "outside." Climate, landscape topography, and geology influence plants and animals and human culture, but not vice versa (at least for the most part). However, many aspects of the environment are strongly influenced by evolutionary systems themselves. For example, humans and other animals extract resources, such as food, from their environments, thereby modifying those environments in ways that influence their own evolution. In the extreme, there might be hardly any distinction between the environment and the system itself: The state of the evolutionary system—i.e., the prevalence of different types of species, music styles, or fashion options—can be the main source of selection determining which types subsequently succeed or fail.[2] This creates strong internal feedbacks and distinct evolutionary dynamics, and so requires a new dial on the Evolutionary Soundboard.

Dial 6: The Frequency-Dependence of Selection

Growing up in the 1970s and 1980s, I knew a lot of boys named Jason or Brian, and a lot of girls named Lisa or Amy. And this was not a peculiarity of my social circle: All four names were among the top ten assigned to babies in the United States in the 1970s (the Canadian database only goes back to 1991). But by the time my childhood friends started having kids of their own in the 2000s, these names were used only rarely: Among the four names, Jason was the highest ranked in the 2000s, at number fifty. Some of the new names appearing in the top ten in the 2000s, such as Ethan and Madison, hadn't even cracked the top two hundred in the 1970s. Parents choose names for many reasons, but researchers have shown that, on average, rare names tend to increase in frequency while common ones tend to decline. The very fact of being rare appears to make a name attractive to parents. Cultural evolutionists call this "novelty bias" or "anticonformist bias," and we can see it at work in trends and fashion choices of many kinds. Sometimes there is nothing more fatal to a fashion trend than its own success.

In evolutionary models, a frequency is a proportion: Out of all people, the proportion of them with a given name is the *frequency* of that name. When the fitness advantage or disadvantage of a particular trait depends on its frequency in the population, we say there is *frequency-dependent selection*. When the thing evolving (e.g., baby name choices) and the environment it is evolving in (e.g., past baby name choices) become one and the same, some new things can happen that we didn't see with directional selection. Trait prevalence feeds back on itself, with profound consequences for maintaining diversity and for the predictability of evolution. And frequency-dependent selection happens a lot.

One of the most genetically diverse components of the human genome (and in the genome of most vertebrate animals) is called the major histocompatibility complex, which contains genes involved in defense against pathogens. Most pathogens, such as bacteria and viruses, go through tens of thousands of generations during one human lifetime, and so they have tremendous potential to evolve and overcome immune system defenses. If any allele in the human population involved

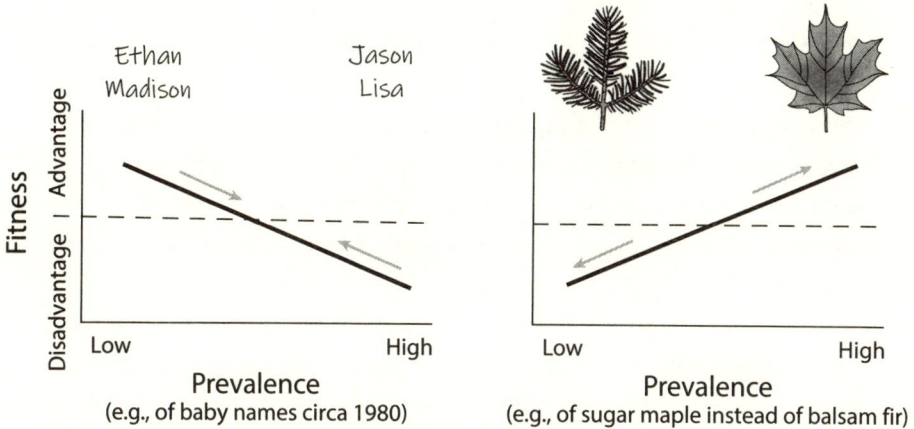

FIGURE 5.4. Frequency-dependent selection: negative (*left*) and positive (*right*).

in defense against pathogens becomes common, pathogens are likely to evolve to overcome its defensive action, such that people with the common allele are more likely to get sick (a selective disadvantage). For a newly generated allele or an allele that has become very rare in the human population, pathogen resistance against its defensive action is much less likely to evolve or persist (e.g., if there is some cost of resistance in the pathogen), such that people with this allele are less likely to get sick (a selective advantage). This process is akin to novelty bias in fashion choices, and it is called *negative* frequency-dependent selection because a high frequency of a given allele is associated with low fitness in individuals with that allele, and vice versa (fig. 5.4, left side). The result is that types (in this case alleles) are protected from being lost, since they gain a fitness advantage when rare, so diversity in the population tends to be very high.

This kind of negative feedback is thought to be a key factor maintaining biological diversity in general. If a species becomes very common, the specific resources it most needs are likely to be depleted (by the species itself), and its predators or pathogens are likely to become abundant. If the species is rare, the resources it needs (in unique ways compared to other species) might be abundant, and enemies rare. Taken together, these dynamics prevent any one species from attaining complete dominance, and they help species bounce back from extremely low abundance. When biologists experimentally increase or decrease the abundances of plant or animal species, they often find some degree of negative frequency dependence like this, stabilizing the diversity and composition of ecological communities. Monocultures, such as we see in many crop fields, are extremely rare in nature and require huge energy investments to maintain (e.g., with herbicides and pesticides).

The flipside of negative frequency-dependent selection is *positive* frequency-dependent selection, under which each type has a disproportionate advantage when its prevalence is high, and vice versa. This form of selection appears to apply in many cultural situations, such as technologies where individuals gain from using the dominant type (e.g., word-processing software), cultural preferences such as those in music, or the prevalence of synonymous words. We have little basis to predict what word processor, song, or word (e.g., sweater versus jumper) will come to dominate, apart from which manages to get a head start.

Most plants depend on root-penetrating mycorrhizal fungi for their acquisition of soil nutrients, with some differences among species in which particular fungi they have this mutualistic association with. In return for nutrients, the fungi receive carbon acquired by the plant via photosynthesis. If a plant species is very rare, the quantity of the mycorrhizal fungi it needs might be insufficient in the soil, in which case the fact of being rare would be a selective disadvantage. Let's say this first species is the sugar maple tree (the source of maple syrup). If the frequency of sugar maple could somehow increase across a threshold—the "tipping point"—a high abundance of its mycorrhizal fungi might confer a selective advantage to the species. The same might be true of a second species—balsam fir—in the same place, with different mycorrhizal partners (fig. 5.4, right side). The general lesson is that when fitness is strongly affected by mutualists—other species from which a given species derives some benefit—there is potential for positive frequency-dependent selection.

Unlike the other forms of selection explored so far, positive frequency-dependent selection does not predictably push the system

FIGURE 5.5. The eastern flank of Mont St-Joseph, in Parc national du Mont Mégantic, Québec, Canada. The light color at low elevation is forest dominated by sugar maple, and the dark color at high elevation is forest dominated by balsam fir. The photo was taken in spring, before leaves of the deciduous sugar maple had fully emerged.

toward one equilibrium state. Rather, there are two completely different potential endpoints of the evolutionary trajectory, with the outcome dependent on where we start. If sugar maple starts out more prevalent than balsam fir, selection should push the system toward maple dominance. If fir starts out most prevalent, selection should push toward a fir monoculture. In my part of the world (southern Québec, Canada), this exact type of positive feedback is thought to maintain the sharp mountainside boundary between forests dominated by sugar maple (at low elevations) and balsam fir (at high elevations) (fig. 5.5). Halfway up the mountainside, climatic conditions are suitable for either species, but we rarely find them in an even mixture.

Selection with positive feedbacks and tipping points of this nature can have potentially profound consequences—for example, the thriving or collapse of an economy or ecosystem. As such, researchers across the natural and social sciences are keenly interested in identifying systems with potentially dangerous tipping points. These will be explored further in chapter 8.

As with the fitness function, the relationship between fitness and prevalence (frequency) can take any shape at all. For example, a given species might be at a disadvantage when extremely rare (because each individual has trouble finding a mate) *and* when highly abundant (because they've used up their resources), with the greatest advantage somewhere in between. Nonetheless, as a first cut, we can add one dial to the Evolutionary Soundboard characterizing frequency dependence, varying between negative and positive.

Selection Is Complex . . . but Simple

The beauty and universality of the idea of selection stem from its focus on the consequences of fitness differences rather than their causes. We could easily fill this entire book (and more) with a list of all the possible causes of fitness differences. It would have to include all the physical and chemical attributes of any location on land, in the sea, or in the atmosphere, all the attributes of all the organisms on which some other organism depends or is otherwise influenced by, and every social, technological, or economic attribute of every human population on Earth. No one wants to do that project.

Researchers have frequently created categories of causes of fitness differences for special recognition, such as human intentions (artificial selection) or direct competition between sperm (transmission bias). But ultimately, these causes, or any others, matter for evolution only to the extent that they turn dials on the Evolutionary Soundboard. For example, crop breeders implementing artificial selection might select only the plants in the top 10% of seed yield or pathogen resistance, with the other 90% contributing nothing at all to the next generation. This creates a fitness function where all but extreme trait values have a fitness of zero, but once we know that fact, we can just plug the relevant fitness function into our evolutionary analysis and proceed as we would with any other case.

In some formulations of the three-ingredient recipe for an evolutionary process, the third ingredient—differential success—is simply called selection. If there were nothing more to differential success than selection, this chapter would mark the end of our treatment of this essential process in the Second Science. Alas, there is more to the story. Some advantages are not systematic, they are random.

6

Randomness and Movement

The island of Tasmania is separated from southeastern Australia by the Bass Strait, a 250-kilometer span of water containing a smattering of smaller islands. Hopping from one little island to the next, you would rarely need to travel on the water for much more than 30 kilometers at a time to get to Tasmania from the Australian mainland. One might thus expect that the Indigenous people of Tasmania had a traditional way of life much like that of the people on the mainland. But when European explorers first visited Tasmania in the late eighteenth century, they were in for a big surprise.

Compared to their counterparts on the Australian mainland, Indigenous people in Tasmania were found to be using a set of tools and technologies that were astonishingly modest in their sophistication and diversity. Indigenous people on mainland Australia possessed hundreds of types of tools and vessels for fishing, hunting, eating, woodworking, making clothes, and transportation. Many of their individual tools were complex assemblages of multiple parts, including tree-bark canoes sewn with plant fibers or animal sinew, and barbed spears made of wood, bone, resin, and twine. In comparison, the people on Tasmania hunted with simple rocks, clubs, or one-piece spears, traveled (not very far) on rafts without paddles, and kept themselves warm by applying grease to their skin and wearing wallaby skins like shawls. They had no tools made of bone, and they did not exploit the abundant fish in the coastal waters.

As archaeologists pieced together the history of the region, the mystery of why there would be such stark differences between two near-adjacent human populations deepened. The toolkit in Tasmania was not always so simple. People have lived on Tasmania for more than thirty thousand years, first arriving when it was not yet an island. At a time when sea levels were lower, mainland Australia was connected to Tasmania by a land bridge. When sea levels rose at the end of the last glacial period, ten to twelve thousand years ago, the land bridge was severed, isolating the population on the island. During the millennia prior to isolation, surviving the glacial cold would have required more than greased skin and loose garments, and archaeological records show evidence of fish consumption and abundant tools made of bone. None of those things remained in use in the eighteenth century. In short, it appears that isolation of this small island population prompted the loss of cultural adaptations and diversity.

Biologists are well acquainted with similar cases in which diversity and adaptations are reduced in small, isolated populations and ecosystems. An isolated oceanic island of a given area typically harbors many fewer species than an equal-sized portion of the closest large land mass. In Melanesia, an island of 100 km^2 contains roughly ten times fewer species of ants than a 100 km^2 portion of New Guinea (the total area of which is $> 750,000 \text{ km}^2$). Small populations of plants and animals typically have less genetic diversity than large ones.

For humans, rare diseases are surprisingly common in small, isolated communities. In these cases, isolation is not necessarily defined by geography, but by a small pool of potential reproductive mates. In the Pingelap Atoll in the Pacific Ocean, roughly halfway between Hawaii and Indonesia, a typhoon in 1775 left a very small number of survivors. The three thousand or so contemporary descendants of these survivors have a 10% incidence of an otherwise extremely rare condition called achromatopsia—a visual impairment involving complete colorblindness. Improbable things seem to happen on small islands, whether those islands are real or virtual.

These examples, and many others, suggest that there is something about the fact of an evolutionary system being small and isolated that reduces diversity and hinders the processes of selection and adaptation. A central aim of this chapter is to explain exactly why that is the case.

In the previous chapter, we saw that selection—systematic differential success among entities based on their traits—is expected to cause a predictable march of trait values from those with low fitness to those of high fitness. One of the main consequences of small size is to make this march less predictable. With small size comes an increasing importance of stochastic events, such that evolutionary trajectories can start to look less like steadily marching toward higher fitness and more like randomly drifting between different trait values.

Drift

Loss of Diversity and Compromised Adaptation

Imagine you have just planted a small section of garden with ten tomato seedlings of two varieties: five beefsteak and five Roma. You plan to use the seeds from this year's harvest for next year's garden. A major hailstorm hits, killing all but two of the seedlings. What are the chances you will have only one variety of tomato to harvest at the end of the summer?

Since hailstones can't aim systematically for beefsteaks (or Romas), let's assume that the survivors are a random sample of the initial ten. Because you started out with a 50:50 ratio of the two types, our scenario essentially involves flipping coins. Each survivor is one coin flip: heads for beefsteak, tails for Roma. The probability of heads is 0.5, so the probability of two flips both being heads is $0.5 \times 0.5 = 0.25$. Likewise, the probability of two tails is 0.25, or 25%. So, the answer to our question is that we have a 50% chance of being left with two plants of the same variety, either beefsteak (heads) or Roma (tails) (fig. 6.1). If this is the entire basis of the future tomato population in your garden, there is a big risk that one storm will leave you stuck with just one variety for a very long time.

Now imagine that instead of starting with 10 seedlings, you start with 1,000. Again, only 2 out of every 10 seedlings will survive the hailstorm, but this time the actual number of survivors is 200. If we flip a coin 200 times, the odds of them all coming up either heads or tails is roughly

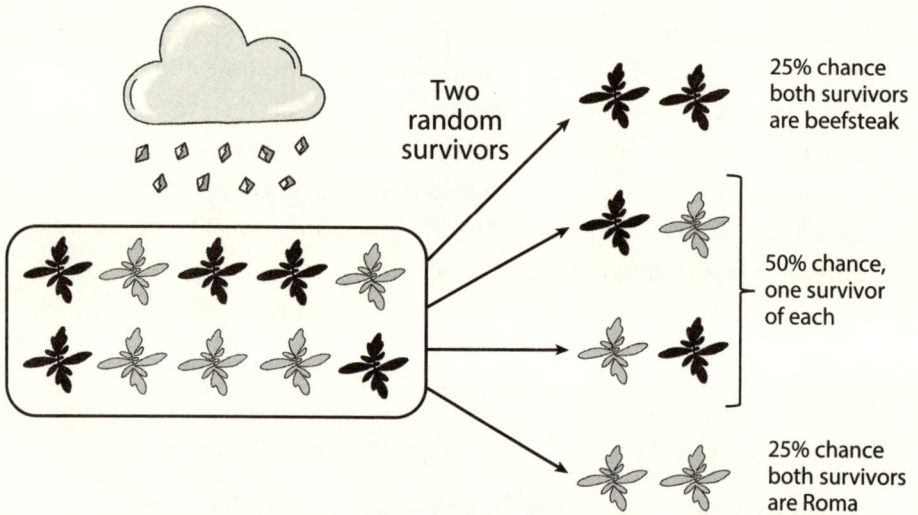

FIGURE 6.1. An illustration of drift in a small population of tomato plants with two types: Roma (gray) and beefsteak (black).

10^{-60}, a number so extremely small that even if the trial had been done every year during the 4.5 billion years of Earth's existence, we wouldn't expect all heads or all tails to happen even once.[1] In short, even after 80% of 1,000 seedlings die, we are effectively guaranteed to still have both varieties of tomato. In this scenario, everything is identical to the first scenario, except the starting population size. By going from 10 to 1,000 seedlings, we reduced the risk of immediate diversity loss from 50% to essentially zero.

The idea of selection, covered in chapter 5, was all about systematic advantages determining which types succeed or fail, which seems incompatible with the idea of life and death as flipping coins. In fact, both can happen at once: Evolutionary systems can involve both deterministic and stochastic processes at the same time. Let's say beefsteaks are more likely than Romas to survive storms. This means that the storm is an agent of selection, providing an advantage to beefsteaks. However, a selective advantage almost never means that there is 100% certainty about precisely what will happen. For example, if selection favors the Omicron variant of the virus causing Covid-19, it means that someone infected with Omicron is more likely *on average* than someone with Delta to transmit the virus to an uninfected person, but one specific person with Omicron might transmit it to no one, while someone with Delta infects ten other people. Similarly, a beefsteak seedling might be more *likely* to survive a storm than a Roma seedling, but this scenario also involves unavoidable probabilities. In essence, we can still view the situation as flipping coins—it's just that now the coins are biased.

Imagine that beefsteaks have a fitness advantage over Romas. For each surviving seedling, we now have virtual coin flips that are more likely to come up heads (beefsteak) than tails (Roma). Let's say the probabilities of heads and tails (which must add to one) are 0.55 and 0.45, which translates to a fitness advantage of 22%, since the difference between the two $(0.55 - 0.45 = 0.1)$ is 22% of the lower one (0.45). In nature, this would be a large fitness advantage, and so we expect beefsteaks to march steadily to dominance. However, if we walk through the same calculations as we did with our unbiased coin, we find that with only two survivors, there is actually a 20% chance $(0.45 \times 0.45 = 0.2)$ that both survivors are Romas—the *disadvantaged* type. The odds of the expected outcome, an increase in beefsteak prevalence, are only 30%. In short, even when selection heavily favors one type over another, the inherent randomness in evolutionary systems can make the actual outcome unpredictable. The tendency for the prevalence of types to bounce up and down at random like this is called *drift*.

In larger systems, the predicted outcome of selection becomes much more likely. If we use a computer to simulate the survival of 200 out of 1,000 seedlings, the 22% advantage leads to an 88% chance that more than half the survivors will be beefsteaks, after just one storm. If we repeat the process every year, taking seeds from the survivors, planting 1,000 new seedlings, and applying the 22% selective advantage, it is virtually guaranteed that within twenty to thirty years beefsteaks will come to dominate, squeezing out Romas completely. This guarantee is lost not only when the population is much smaller (as we have seen), but also when the selective advantage is weaker. In other words, drift and selection interact. In a population of many thousands—a common occurrence in nature, especially for small organisms—even a selective

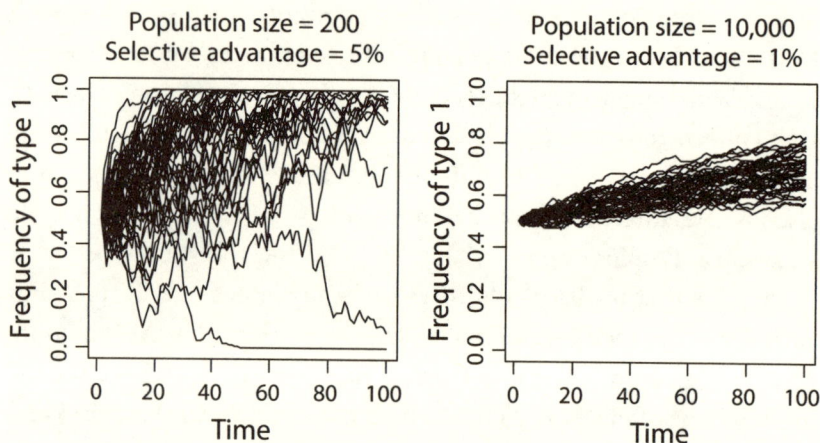

FIGURE 6.2. Changing frequencies of one hypothetical type of entity in competition with a second, during fifty computer simulations. When population size is relatively small (200, *left*), even with a relatively large selective advantage (5%), the favored type 1 sometimes goes extinct. With a large population size (10,000, *right*), type 1 is virtually guaranteed to increase in frequency, even with a small advantage (1%).

advantage of 1% is almost certain to lead to eventual dominance of the advantaged type. For many combinations of small population size and/ or weak selection, there can still be plenty of drift, even when selection creates a tendency for certain types to increase or decrease in prevalence (fig. 6.2). A tendency is not a guarantee.

These tomato-garden scenarios can help us understand why, in general, small evolutionary systems tend to be less diverse and also less likely to retain adaptive characteristics. On a small oceanic island, the inescapable probabilities involved with birth and death can elevate the rate of species extinction and the rate at which alleles are lost.[2] The fact that many of the lost types might have actually been those at a selective advantage means that the process of adaptation is slowed down. If all island denizens adapt relatively slowly, the playing field *on the island* is even, and so many species nonetheless thrive, often evolving unique characteristics (e.g., large flightless birds). On continents, where population sizes are much larger, small selective advantages should be able to sweep through populations more easily, accumulating over the long term to produce much more competitive organisms. It is thus

not surprising that when previously isolated biotas become connected—
either by the formation of land bridges, by opened waterways, or by
people moving organisms around—the species from the larger place
often take over the smaller place. This is especially striking following the
introduction to new regions of nonnative plants and animals: In many
parts of the world, introduced rats, cats, pigs, snakes, toads, and plants
of many kinds have had their most dramatic effects on islands. Between
continents and islands, the playing field is not even.

Having covered the essentials of how drift works in evolutionary systems, we can return to Tasmania and its historically isolated Indigenous
population. The role of drift in cultural evolution is not as simple as
flipping coins, but the core underlying principles are the same. In a
small, isolated group of people, we expect strong constraints on diversity and adaptation due to a magnified importance of randomness. The
following account is drawn from the work of cultural evolutionist Joseph Henrich, who has combined mathematical models and observations of small, isolated human communities—including Tasmania—to
provide a compelling account of cultural drift in action.

The cultural knowledge required to build a sophisticated object like
an airplane or a sea-faring canoe is rarely housed in the brains of individual people, but rather in a society's "collective brain." Some people
might know how to acquire raw materials, others how to process them
for use in building, and still others how to assemble them into a working
transportation vessel. Every aspect of the process might be under selection based on how well different materials or methods meet the demands people have of planes and boats. When a large, well-connected
population becomes small and isolated (e.g., due to a disease outbreak),
two things can happen: First, the people that survive the process might
not include those with the deepest knowledge of one or more aspects
of building something from scratch. This is like losing from our garden
a tomato variety that was the best for some medicinal concoction. Second, the process of learning across generations, which is necessary for
cumulative and adaptive cultural evolution, becomes less efficient. This
second point requires some explanation.

Imagine a point in time at which the dominant design of canoe is well
suited to people's needs (sea travel, fishing, transportation), in which

case most deviations from the design would make it worse. Because students rarely succeed in perfectly imitating the master builders, most canoes that students build on their own will not be as good as the model they were emulating. In a large population, we might expect a few students to attain the skill and knowledge of the masters. Some might even do a bit better. With a great many judges of the final products—not only builders, but users as well—subtle improvements or defects will not go unnoticed. Finally, with a strong communication and cooperation network, improvements can spread quickly, defects can be flagged, and new masters can guide others to ensure the quality standards of the next generation of canoes. Slowly but surely, designs are likely to improve.

In a small population, we still have master builders and students, but important aspects of learning in the collective brain are compromised. The odds that there will be any students who equal or exceed their masters' expertise are reduced, as is the likelihood that minor improvements or defects will be detected. Larger differences would presumably be noticed, but over the long term, major improvements often result from an accumulation of many little ones, and even maintaining the status quo depends on weeding out designs with slight deficiencies. If the little things go undetected, quality eventually deteriorates. Just as we saw in the tomato garden, the long-term expectation here is a loss of diversity and a decreased likelihood of generating and maintaining adaptive variants. And the reason is essentially the same as it was in the garden: small selective (dis)advantages are overwhelmed by random aspects of change and inheritance. In a sense, cultural adaptation involves crowdsourcing knowledge and expertise; this works well in a big crowd, but not so well in a small one.

To see with my own eyes the influence of imperfect information transmission (learning) on adaptive cultural evolution, I conducted a little experiment using an instruction for tying shoelaces. I started with a sentence taken from a set of online instructions: "Form a loop with one end of the shoelace by bringing the middle of the loose segment to the base of the overhead knot." If this instruction is passed down orally, we can imagine the learning process as being akin to translation—each person hears the instruction in their own way, via the filter of their unique brain. To simulate this transmission process, I used an online

Original Instruction: Form a loop with one end of the shoelace by bringing the middle of the loose segment to the base of the overhead knot.

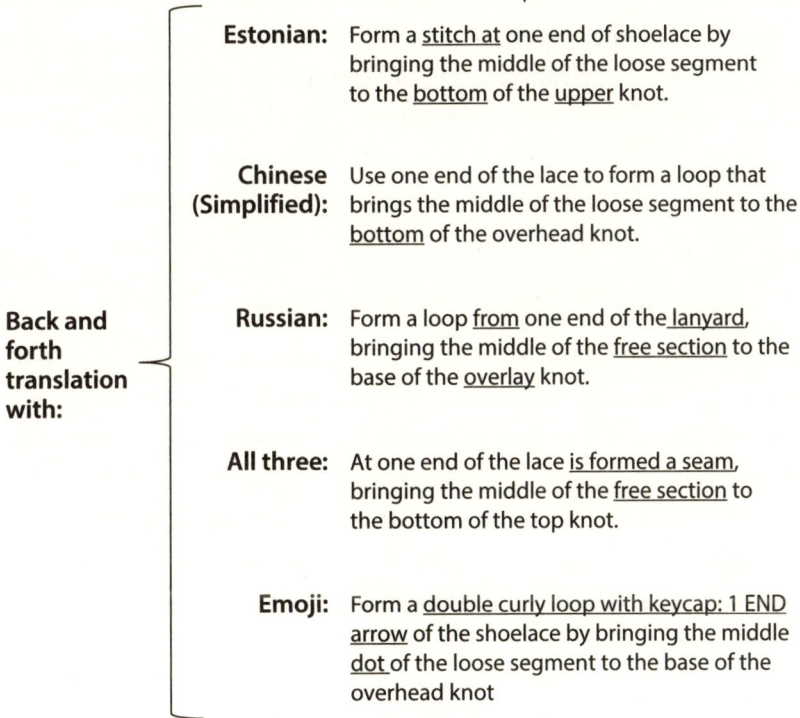

Estonian: Form a <u>stitch at</u> one end of shoelace by bringing the middle of the loose segment to the <u>bottom</u> of the <u>upper</u> knot.

Chinese (Simplified): Use one end of the lace to form a loop that brings the middle of the loose segment to the <u>bottom</u> of the overhead knot.

Russian: Form a loop <u>from</u> one end of the <u>lanyard</u>, bringing the middle of the <u>free section</u> to the base of the <u>overlay</u> knot.

All three: At one end of the lace <u>is formed a seam</u>, bringing the middle of the <u>free section</u> to the bottom of the top knot.

Emoji: Form a <u>double curly loop with keycap: 1 END arrow</u> of the shoelace by bringing the middle <u>dot</u> of the loose segment to the base of the overhead knot

Back and forth translation with:

FIGURE 6.3. Results of translating an instruction for one step of tying shoelaces from English to one or more other languages and back. Notable differences are underlined.

translator to translate the sentence into three languages: Estonian, Simplified Chinese, and Russian. In each case, I reversed the direction of translation repeatedly until the English version stabilized (usually after two or three rounds).[3] I then took the final English sentence from the Estonian back-and-forth, cycled it through Chinese and Russian, and then back to English, so the sentence had also been through all three languages. To push the experiment to the limit, I also conducted one back-and-forth translation to and from an emoji-laden version of English. The results are shown in figure 6.3.

Contrived as it is, this example is quite instructive. If a population crash left only emoji-thinking survivors, shoelaces would probably disappear from the culture in short order. Cycling through Estonian and Russian made the instruction harder to understand (for me anyway), while going through Chinese might have produced an even better instruction than the original. Even with these exaggerated learning imperfections, we can glean a key lesson: Improvement is improbable for any one variant (in this case a translation), so it is unlikely to happen in a small population where there are few attempts to learn. A large population is quite likely to hit on variants that lead to improvement.

Exploring New Trait Values Can Facilitate Adaptation

Up until now, our consideration of drift has focused on a single small place or population—an island, either literally or figuratively—for which we often expect a loss of diversity and compromised adaptation. However, there is a sense in which drift can play a *creative* role in evolution, potentially promoting long-term adaptation. To understand how it works, we need to bring back the fitness function from chapter 5, in which fitness was shown to vary according to traits. We will start with a fitness function that has two peaks of different height (fig. 6.4), which could represent boats or birds for which it's better to be small (but not too small) than medium-sized, while being large (but not too large) is best of all. Fitness functions like this show peaks and valleys and so are often called "fitness landscapes" (especially when fitness is shown to vary as a function of two traits simultaneously instead of just one).

Let's say the trait in question is the size of a bird, and that the average bird in a population is relatively small, but not as small as the size corresponding to the lower fitness peak. This is represented by the triangle in figure 6.4. From this starting point, new birds that are slightly smaller than the average will have higher fitness, and so selection will push the population average toward smaller size (the peak on the left). If the population contains thousands or millions of birds, the uphill motion will be highly predictable—unavoidable really, since selection makes any downhill movement essentially impossible. Once at the peak of the

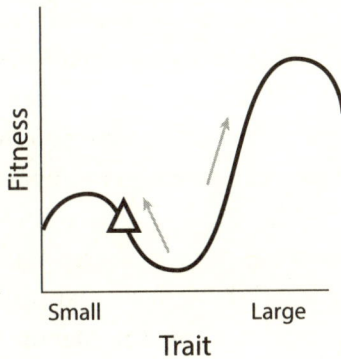

FIGURE 6.4. A hypothetical fitness landscape: a relationship between fitness and traits that has peaks and valleys. The triangle is a hypothetical starting point for subsequent evolutionary change.

small hill, the population should be stuck there. The average size of birds will stay small indefinitely.

Since most new variants are only slightly different from their predecessors (i.e., movements left or right in fig. 6.4 cover short distances), we can assume that it is impossible for a new variant to suddenly appear on the taller hill. A sparrow-sized bird (at the lower peak on the left) can't give birth to a duck-sized offspring (near the higher peak on the right). But imagine that the population declines precipitously in numbers. In a population with just a few individuals, the average size of bird may well drift upward, against the force of selection, across the fitness valley, and toward the size of a duck. In general terms, random drift makes it more likely for the average trait value—whether of birds, boats, or anything else—to move in any direction, even if it goes against the direction favored by selection. If the population finds itself somewhere on the big hill and then rebounds in numbers, selection could then push it to the top. Without drift, selection would have actually *prevented* the population from achieving the trait value with the highest fitness, by keeping it stuck on the small hill. While a population decline runs the risk of outright extinction, if the population survives, drift will have allowed it to explore new and different options that might otherwise have never been explored.

Random changes in the frequencies of different types might also push an evolutionary system over a threshold, after which positive frequency dependence can kick in, leading one type to dominance (see fig. 5.4, right side). Take the case of synonymous words (e.g., lorry versus truck), for which there is evidence for positive frequency-dependent selection. This means that there is a communication advantage if everyone uses the same word (e.g., for a large road vehicle used for the transport of goods or people), but it doesn't really matter which one of the two they all decide to use. Drift can enter the equation by allowing the word "lorry" to randomly achieve a higher prevalence than the word truck, after which the positive feedback kicks in, and so lorry it will be, for everyone, from then on. It just as easily could have been truck.

The essential lesson about drift is that there are probabilities inherent in the events driving evolutionary change, and that these probabilities introduce some randomness into evolution. In a large system, even weak selection can overwhelm the randomness, rendering drift close to powerless. But in a small system, drift can have a major impact. Drift will typically reduce diversity and impede adaptation, although in special circumstances drift can give adaptation a boost by allowing trait values to test out distant parts of the fitness landscape. One important condition for all these lessons is that they apply to small systems that are *isolated from others*. Once that isolation is broken, we need to consider how much connection there is between each subsystem and the whole of which they are a part (e.g., tomato gardens across a neighborhood, islands in an archipelago, or national economies within a global economy). In other words, we need to think about movement.

Movement

If stories of population isolation reveal the influence of drift, stories of reconnection can tell us a lot about the consequences of movement. An especially revealing case comes from the northernmost outpost of humanity on Earth, in northwestern Greenland. There, the Polar Inuit live in coastal areas that are covered in snow and ice for most of the year. An epidemic that hit the population in the 1820s killed many of the people with the deepest knowledge of how to produce critically important tools, such as bows and arrows, complex spears called leisters (with multiple prongs and barbs angled backward), and kayaks. These technologies presumably evolved via many cumulative steps over the generations, with the community's collective brain storing the knowledge, which could not be easily recovered once lost. For several decades, with a diminished ability to hunt caribou and seals, or to catch fish, and no means of travel to contact Inuit elsewhere, the population declined.

In the 1860s a group of Inuit from nearby Baffin Island arrived in Greenland with their kayaks and tools in hand, along with the knowledge of how to make them. These tools were rapidly reintroduced into the local culture, and the population rebounded. At first the Greenland kayaks more closely resembled those of Baffin Island (relatively large) than those of their ancestors (smaller), but over a period of decades, the design gradually converged on the original local version. In a sense, the population was rescued by the incoming movement of knowledge and ideas, carried back to Greenland by the people from Baffin Island.

Small, isolated populations of plants and animals can also be rescued by incoming movement (i.e., immigration). If we head about 7,000 kilometers southeast of northwestern Greenland, across the Arctic, we get to Jämtland County, Sweden, where researchers have studied a small, isolated population of Arctic foxes. Intensive hunting for fox fur in the late nineteenth century caused a massive decline of the species in Scandinavia, from many thousands of foxes to a few hundred. Genetic analysis of this one population in Jämtland County indicated that it was founded by fewer than ten individuals. Just as a student of canoe building has few people to learn from in a small population, a fox in Jämtland has few potential mates. Following the population for nine years from 2000 to 2009 revealed complete isolation from other populations, intense inbreeding, and persistently small size. But things were soon to change. The arrival of three "outside" males in 2010 was quickly followed by marked increases in breeding success, survival of young, and genetic diversity. Five years later the population had doubled in number. Similar examples of "genetic rescue" have been observed in various plants and animals, including panthers, bighorn sheep, pygmy possums, guppies, trout, and prairie chickens.

These cases, in which movement countered the detrimental effects of drift and inbreeding, raise an important question, first encountered in chapter 3: How do we draw a line around a population, or, more generally, around what we have been calling an evolutionary system? Both examples described so far in this chapter involved a "here" (Greenland or Jämtland County) and a "there" (Baffin Island or the rest of Scandinavia). But how do we decide what counts as here and there? The fact that cultural evolution led the Inuit of Greenland and of Baffin Island to different kayak designs, one of which was lost for a while, tells us that there is some meaningful distinction to be drawn between the two populations. Scandinavian Arctic foxes almost always find mates within their local population, and only rarely by mixing with foxes from elsewhere, which makes those population distinctions meaningful as well. But Greenland Inuit and Jämtland Arctic foxes can also be considered as part of larger evolutionary systems. Different Inuit communities and different Scandinavian Arctic fox populations interact more often with one another than either does with, let's say, the Indigenous people of Australia or Arctic foxes in Alaska, respectively. We could also say that all of humanity (*Homo sapiens*) is an evolutionary system, as is the global population of Arctic foxes (*Vulpes lagopus*).

Another way to approach the question of drawing boundaries is to ask if it was movement that rescued these populations, or if it was the fact of their becoming part of a much larger population. Did these small, isolated populations become connected to others, or did they get subsumed into a larger whole? The answer is: both. When the European Union was formed in the early 1990s, it meant greater flows of goods, services, and labor across national borders, *and* it meant that each member country was now part of a new, larger political whole. When the Suez Canal opened in 1869, many plants and animals were allowed to move between the Mediterranean Sea and the Red Sea, *and* the two became part of one larger body of salt water. These are just two ways of viewing the same thing. The size of a system—and so the importance of drift—is intimately linked with the degree of movement between systems, or between the parts of one system.

To illustrate the evolutionary consequences of movement, let's bring back our pie charts representing types of slime molds, companies in an industry, competing hypotheses, or any other set of variants in an evolutionary system. The process of variation generation was described in chapter 3 as occurring *within* a system—a mutation in a slime mold, a spin-off company, or a new twist on an old hypothesis. In many real evolutionary systems, these internal sources of variation are actually of modest importance compared to variation that enters from the outside. Spores of a different slime mold blow onto a plate from elsewhere; a company from the software sector enters the online data-storage industry; a hypothesis from economics is imported into biology. Movement is of immense importance in the contemporary world: Within the borders of many countries like the one I live in (Canada), much of the current cultural diversity, and a good chunk of the biological diversity, has been introduced from elsewhere, via immigrating people or via the things those people have brought along with them, deliberately or inadvertently. Of the plants and animals that make up my daily diet, most are not native to Canada, including wheat, rice, potatoes, tomatoes, carrots, bananas, and chickens. In short, if we assume that the variants coming in from the outside are different than those on the inside, movement represents an important source of variation and so should augment local diversity (fig. 6.5).

Movement can have a strong influence on the importance of drift, and on the raw material required for selection to act. If the "everywhere else" depicted in figure 6.5 is much larger than our focal system (e.g., a continent near an island), it can act as a constant source of new variation, continuously reintroducing types that might be lost to drift. Movement counters the effects of drift. At the same time, by augmenting diversity and thus the range of options to select from, movement can also promote adaptation. However, if the types coming in are less well adapted to local conditions (e.g., polar bears swim to the tropics),

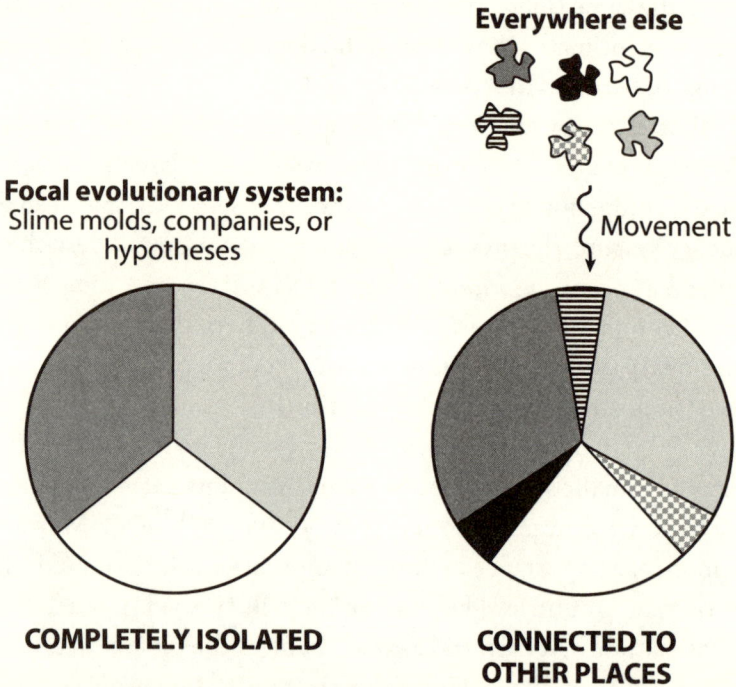

FIGURE 6.5. Movement into an evolutionary system can increase local diversity.

they might just take up space that would otherwise be occupied by the well-adapted types, with the overall effect of impeding adaptation.

Instead of thinking of movement into one system from a vaguely defined "everywhere else," some situations—like an archipelago of islands—force us to consider multiple systems of comparable size. With this shift in perspective, one additional consequence of movement becomes evident: homogenization. If we consider two islands that evolved under different conditions, we might expect their cultures or biotas to be quite different (fig. 6.6, left side). If we now connect the two islands by allowing frequent movement between them, perhaps the original types remain most prevalent locally, but the establishment of new types in both directions will have two consequences (fig. 6.6, right side). First, as we already saw, diversity will increase locally. Second, the two places will come to resemble each other—that is, they will become homogenized

TWO COMPLETELY ISOLATED SYSTEMS

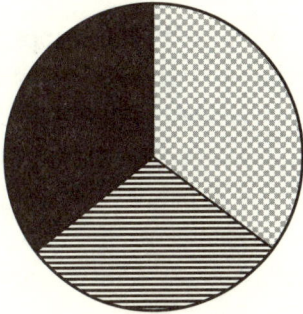

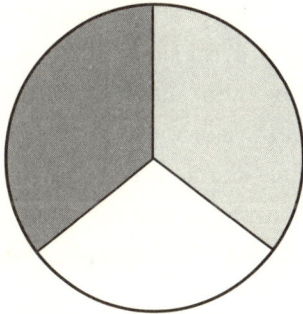

TWO SYSTEMS CONNECTED BY MOVEMENT

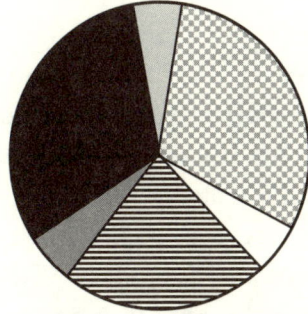

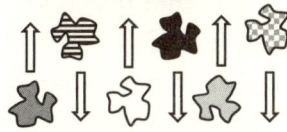

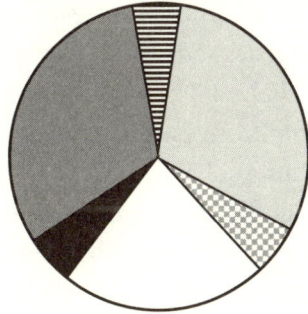

FIGURE 6.6. Movement between systems makes them more similar—it homogenizes them.

to some degree. We can see this in the many aspects of the world today that have been homogenized due to massively increased human movement from place to place. Biologists call this "biotic homogenization," while social scientists use a variety of terms, including "cultural convergence," "cultural globalization," and "McDonaldization." To the unique biological and cultural elements found in different regions of the Earth, humans have added a bunch of common elements to many of them, including strip malls, fast-food restaurants, rats, pigeons, and dandelions. There will be much more to say on this topic in chapter 9.

Completing the Evolutionary Soundboard

Dials 7 and 8: Size and Movement

Characterizing drift and movement require two final dials to complete the Evolutionary Soundboard (fig. 6.7). For the ultimate determinant of the importance of drift, we need the size of the evolutionary system. In cases where we can easily count entities (e.g., alleles, people, businesses), this is essentially population size. In other cases, it might be more appropriate to assess size in terms of variables like land area, biomass, or gross domestic product. All else equal, a small island or country cannot support the same quantity of life and its products as a large one, and we can tune this dial to capture that difference. Because small size enhances drift, which is the random component of differential success, we place this dial in the "Differential Success" portion of the soundboard.

The second dial is the rate of movement, of whatever is being moved, from genes and organisms to products and ideas. In the simplest scenario, this dial might characterize the rate of movement into one isolated system from everywhere else. More complex scenarios might involve multiple subsystems, each with its own dials characterizing internal processes, and a dial characterizing movement from one to the other. Again, for the sake of keeping things simple, movement is represented by a single dial.

The Evolutionary Soundboard is now complete. This is, of course, a simplified version of what we would need to build an actual mathematical model of an evolutionary system, in the sense that there are many complexities one could add. Evolutionary systems present no end of unique features. Our attention is frequently drawn to system-specific factors that underlie variation generation, selection, or movement, including everything from storms, radioactivity, and limited resource availability to musical tastes, advances in technology, and political unrest. From the view of the Second Science, the items on this potentially endless list matter only to the extent that they cause one or more of the dials on the Evolutionary Soundboard to be tuned up or down.

In its generalized form, evolution has been called an algorithm that applies to any system that meets a few criteria (see chapter 3). The

THE EVOLUTIONARY SOUNDBOARD

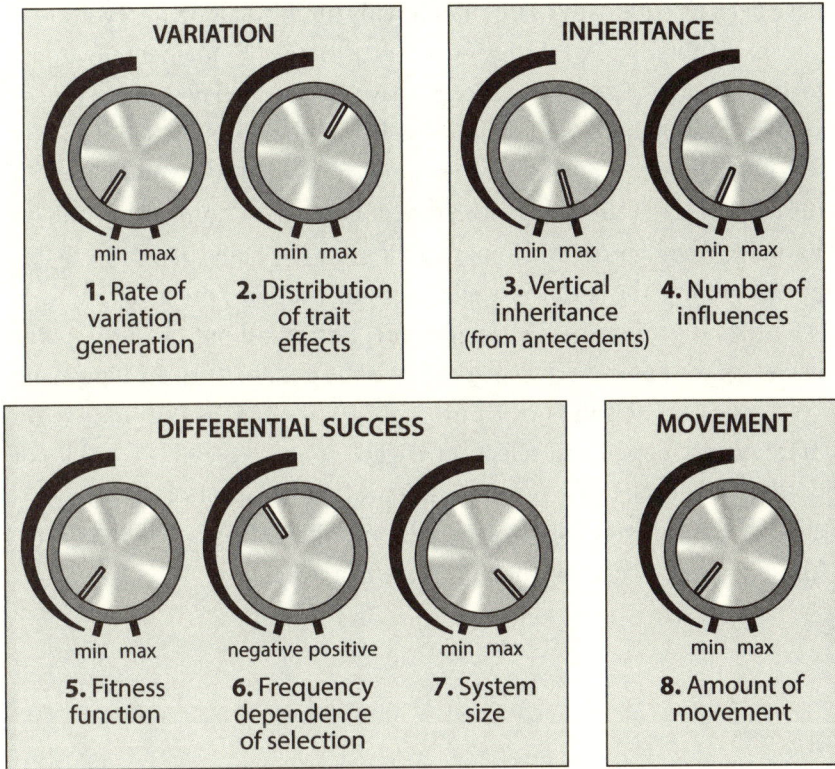

VARIATION

min max   min max

1. Rate of   **2.** Distribution
variation   of trait
generation   effects

INHERITANCE

min max   min max

3. Vertical   **4.** Number of
inheritance   influences
(from antecedents)

DIFFERENTIAL SUCCESS

min max   negative positive   min max

5. Fitness   **6.** Frequency   **7.** System
function   dependence   size
                  of selection

MOVEMENT

min max

8. Amount of
movement

FIGURE 6.7. The complete Evolutionary Soundboard. Dials 1–6 are set as in previous chapters. Dials 7 and 8 are set arbitrarily to represent large systems with only a small amount of movement between them.

Evolutionary Soundboard captures the universal features of this algorithm. The number of causal factors lurking in the background is mind-boggling, but the range of possible evolutionary outcomes is beautifully succinct: The prevalence of different types in an evolutionary system will change only if new types are produced locally or arrive from elsewhere, or if the existing types have differential success, because of either systemic (dis)advantages or random accidents.

Having established the fundamental principles of the Second Science, we are now poised to delve more deeply into the many ways that understanding and applying evolutionary ideas are of immense importance to humanity. This importance takes numerous forms. For millennia, people have been intentionally or unintentionally tuning dials on the evolutionary soundboards of wild species and ecosystems, fundamentally transforming them in the process. People have also created new evolutionary systems from scratch. Culture is the main one already discussed at length. The next frontier is evolutionary systems housed inside computers—in the form of artificial intelligence. In recent centuries, human activities on Earth have had a profound impact, giving rise to a name dedicated to this period of time: the Anthropocene.[4] Among the many preoccupations of scientists regarding the Anthropocene, two stand out from the point of view of the Second Science: The first is the possibility of pushing many systems past the kinds of tipping points we saw in the discussion of frequency-dependent selection, triggering unpredictable cascades of positive feedbacks. The second is the profound impact of massively increased movement of species, people, and their ideas across the Earth. These topics will occupy the next three chapters.

PART III

The Second
Science in Action

7

Harnessing and Directing Evolution

On February 16, 2023, I created an account with the company OpenAI so I could test out their chatbot, ChatGPT. A chatbot is meant to simulate human-like conversation with its users—in this case, me. Here was our first exchange:

> ME: *Do you think the word evolutionary is appropriate for describing the ChatGPT algorithm?*
>
> CHATGPT: *Yes, I think the word "evolutionary" is appropriate for describing the ChatGPT algorithm in the sense that it has been developed through an iterative process of training and refinement that mimics the process of biological evolution.*
>
> *The ChatGPT algorithm is a type of language model that has been trained on large datasets of text using unsupervised learning techniques. During training, the model learns to recognize patterns and relationships in the data, and gradually becomes more accurate and sophisticated in its ability to generate human-like text.*
>
> *This process of training and refinement is similar to the process of biological evolution, in which organisms adapt and evolve over time to become better suited to their environment. In the case of ChatGPT, the "environment" is the vast corpus of text that the model is trained*

*on, and the "adaptation" is the model's ability to recognize and
generate human-like language patterns.*

*So in summary, I believe that the word "evolutionary" is a fitting
way to describe the ChatGPT algorithm's development and
refinement over time.*

By the time you are reading this book, exchanges like this one between human and machine will probably seem unremarkable, mun-
dane even, given how quickly we seem to transition from amazement
at new technology to taking it for granted. But as I write these words,
I am still firmly in the amazement camp. The entire four-paragraph
answer streamed out within about ten seconds, and if this were an answer to an undergraduate exam question, it would probably get an A.
Indeed, I entered a few questions from past exams in my plant ecology
class—in French—and got perfectly good answers, also in French. The
content was not always 100% accurate, but it was easily above average,
as was the quality of the writing. Almost all students in my class pass
the final exam, which takes them two or three hours to write. ChatGPT
could do it in a few minutes. Notwithstanding reports of factual errors
and bizarre behavior, the quality of answers from ChatGPT is astonishing. How does it do it?

The short answer to this question is that programmers have figured
out how to get computers to *evolve* solutions to difficult problems. This
is a major deviation from the now-abandoned idea that programmers
must always tell a computer what to do in every possible situation. The
result of evolution in silico is artificial intelligence (AI), which is one of
three examples to be discussed in this chapter of how people have har-
nessed or directed the power of evolution toward achieving specific
goals. We will first delve more deeply into the evolutionary process by
which computers achieve AI. Next, we will pick up the story of plant and
animal breeding that was started in chapter 5, bringing it up to the contemporary use of genetic engineering, to direct biological evolution. Finally, we will explore the cultural counterpart of genetic engineering—
"memetic engineering"—essentially, advertising and marketing as tools
used to direct the course of cultural evolution.

The Evolution of Artificial Intelligence

The model underlying ChatGPT is of staggering complexity, but the essential features can be understood by first considering simpler tasks undertaken by machines. From there, we can work back to more complex applications. To begin, consider the task of classifying images of shapes as one of three possibilities: square, circle, or triangle. Let's say we have divided each image into nine pixels, and the input data for attempting classification are the proportions of each pixel occupied by the shape. In the case of the square shown in figure 7.1, each corner pixel (numbers 1, 3, 7, and 9) has a value of 0.25; each side pixel (2, 4, 6, 8), a value of 0.5; and the center pixel (5), a value of 1.0. Compared to the square, the circle has smaller values in the corners and larger values on the sides, while the triangle would have values of zero in the top two corners.

During early attempts to create "intelligent" machines, the idea was that programmers could give computers very specific instructions on how to recognize different shapes, numbers, letters, voices, or faces. In our example of shapes (fig. 7.1), this seems fairly straightforward. Essentially, we can tell the computer which combinations of the nine numbers indicate whether we have a square, a circle, or a triangle. Even if shapes are not perfectly formed, one can imagine a human-written algorithm of if-then steps allowing a computer to recognize different types. In this case, as programmers use the age-old process of trial and error in writing code, the algorithm might evolve, in the same way that canoes, spears, cell phones, or any bit of technology evolves. But rather than *asking* the computer to figure out how to recognize shapes, humans are always *telling* the computer how to do it.

A key step on the path that led to ChatGPT, and to other applications of AI such as facial recognition, self-driving cars, or playing chess, involved outsourcing much of the evolutionary process to the computer itself. Even if we bump up the complexity of a task just a little bit, it becomes extremely difficult for humans to write a fixed set of instructions that will perform well. Recognizing shapes, numbers, or letters drawn by hand, in many different styles, represents a monumental

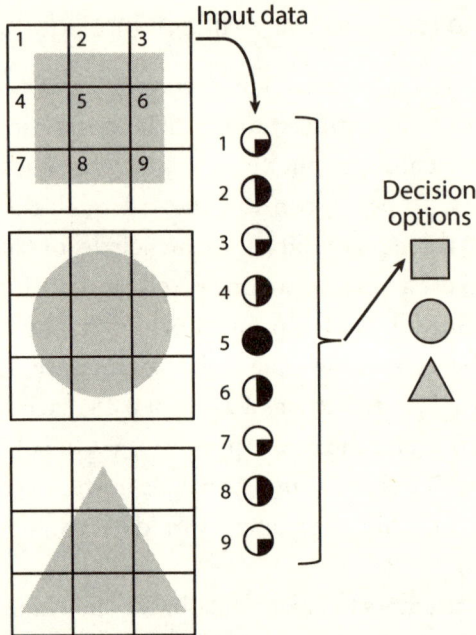

FIGURE 7.1. Simplified analysis of images used to classify shapes based on how much of each of nine pixels is filled by the shape.

challenge, to say nothing of recognizing faces photographed from different angles and under different lighting conditions. And if we couldn't get computers to do those things, ChatGPT would still be a distant dream. Luckily, it turns out that evolution in silico works remarkably well. Computer scientists call it "deep learning," which is now routinely used to ask computers to figure things out on their own.

A key tool for deep learning is called an artificial neural network, designed to mimic (very roughly) the human brain. A simple example is shown in figure 7.2. In the case of shape recognition, we have our nine input variables (pixels), from which we want to make one of three final decisions: square, circle, or triangle (the output). The next steps might sound mysterious to anyone not already well-versed in neural networks, but hopefully by the end of the next paragraph some clarity will be achieved. Between the input data and the decision "layers," programmers insert any number of additional layers of "neurons" that ultimately

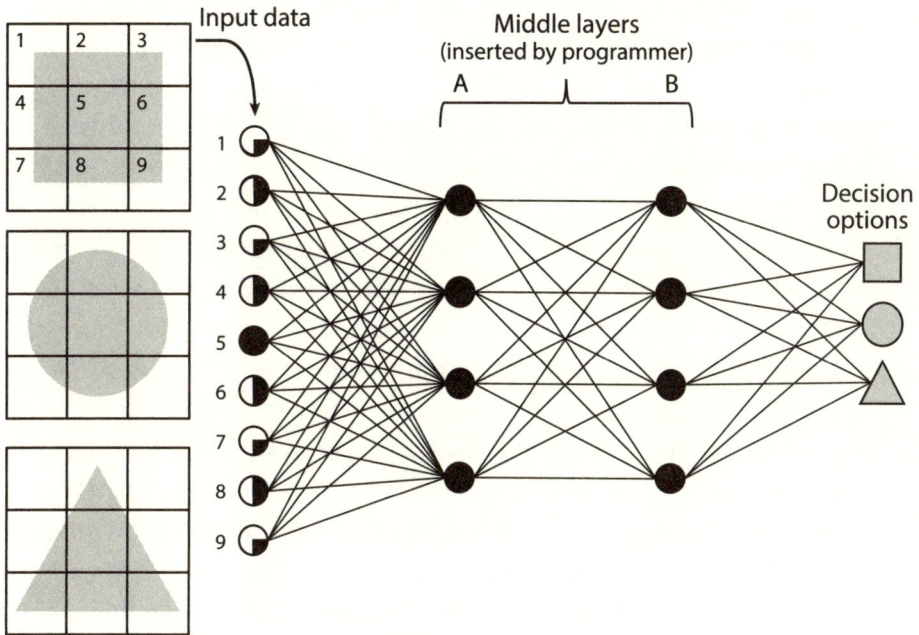

FIGURE 7.2. A neural network that could allow a computer to evolve an algorithm for classifying images of shapes.

relay the information from the input layer to the decision layer.[1] Let's say there are two middle layers, each with four neurons (fig. 7.2). In middle layer A, the input to each neuron consists of the values from each of the nine image pixels, from which an average is calculated to give a single number. The average involves weighting each of the inputs, which means that, for example, pixel 1 might have a big influence on a given neuron in layer A but a small influence on another, and vice versa for pixel 2. Each neuron in middle layer A also has a threshold value to decide whether it will send a "signal" along to layer B. If its weighted mean input exceeds the threshold, it sends a signal; otherwise, it does not. Each neuron in layer B also calculates a weighted average of its inputs, and then, based on its own internal threshold, "decides" whether to send a signal along to the final decision layer. For the three options in the decision layer, a score is calculated just as it was for the mid-layer neurons, and the final decision is based on the one with the highest score.

If you're like me (before I watched some great YouTube videos), the description in the previous paragraph probably makes little sense, because nothing has been said about where the weights and thresholds—so critical to the shape recognition task—come from. This is where evolution comes in. When building a model to recognize shapes, the weights and thresholds are initially assigned random values. This "unintelligent" model is then asked to classify images of shapes that have been preverified by people. Of course, this first model will perform poorly, essentially classifying shapes at random. Using some mathematical trickery, however, the computer can identify what small changes to the initial weights and thresholds would improve model performance from that random starting point. Those changes are then implemented, the model is rerun, and again the nature of the mistakes is used to assess how the weights and thresholds could be tweaked once more to improve things. This process of trial and error—that is, this *evolutionary* process—is continued until no further improvements are possible. At this point, the training phase of model development is complete. Then, a new set of preverified images (i.e., images not used in the training phase) are used to test the model's performance in the validation or testing phase. As more input data become available, the model can be repeatedly retrained and retested in a perpetual evolutionary process.

Another way of viewing the model is as an equation with as many parameters as there are weights and thresholds—basically one for every middle layer neuron (8) and connecting line (64) in figure 7.2. If an equation with 72 parameters sounds complicated, consider that the model underlying ChatGPT has 175 billion parameters (as of February 2023). Pause for a second. That's 175 *billion* parameters, trained using endless terabytes of human-written text. Writing down the ChatGPT model in 12-point font on one line would require a piece of paper that is longer than the distance from the Earth to the moon.[2] Even in highly simplified teaching examples for neural networks, such as number recognition, there is a degree of complexity that makes it near impossible to fully understand *why* the final model looks the way it does. In a popular example, images of numbers are represented by 784 input variables (28×28-pixel images), middle layers have 16 neurons each, and there are 10 possible outputs (the numbers 0 to 9). That makes for an equation with more than 10,000 parameters. In essence, we are looking for the optimal values for all 72, 10,000, or 175 billion parameters simultaneously. There is just no way of telling a computer exactly how to calculate the answer, but an evolutionary algorithm can be written that will allow the computer to find the answer by trial and error.

To relate things back to the evolutionary principles covered in previous chapters, the parameters in a neural network model can be thought of as traits—numbers that characterize the model, the same way that height, weight, and resting heart rate might characterize a lizard or a cheetah (or you or me). The fitness of the model is its performance on a validation dataset: how often it gets the right answer. Finding optimal parameter values is thus one more example in which selection pushes the traits in a system toward a peak in a fitness landscape. The fitness landscapes in previous chapters showed just a single trait, but in reality, fitness landscapes involve many traits. Every one of the 3 billion nucleotides in a human genome could be thought of as a trait, just like the 175 billion parameters in the ChatGPT model.

Unfortunately, our human brains can't visualize more than three dimensions easily, but even with fitness shown as a function of two traits (fig. 7.3), we can illustrate the essential evolutionary principles at play. In building a neural network model, choosing initial parameter values means plopping oneself down at a random spot in the landscape—a random process of generating a new variant. Based on the nature of errors made on a training dataset (e.g., good success at classifying squares but not triangles), mathematical tools can be used to look around the local portion of the landscape, and so to "see" what direction from that starting spot points uphill.[3] Repeated rounds of parameter tweaking (variation generation) and model testing (selection) push the system uphill (adaptation), until eventually no further improvement is possible. We saw the same process at work, for one trait instead of two, in chapters 5 and 6 (e.g., fig. 6.4). Anytime more training data are added, the shape of the landscape will likely change, and the evolutionary algorithm can be reinitiated to find a new peak. For a neural network, instead of moving around in only the three dimensions that we can

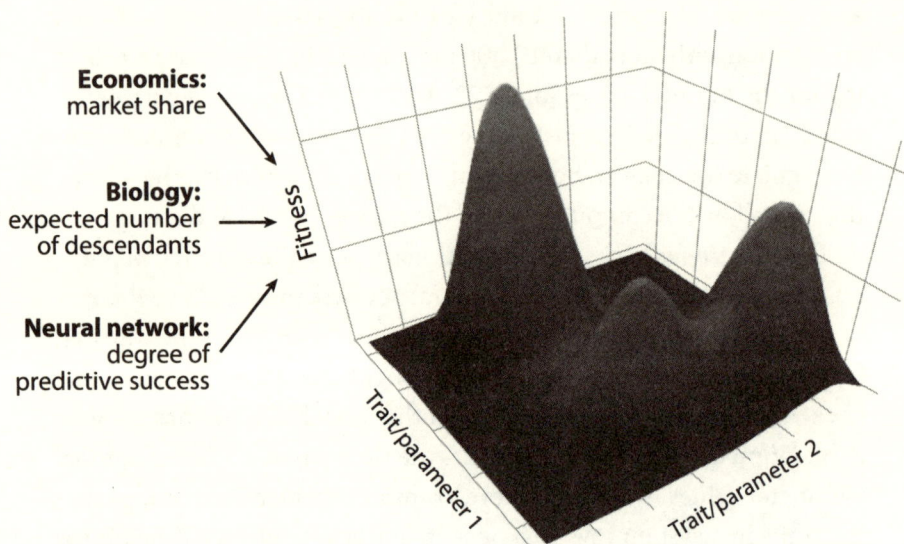

FIGURE 7.3. A hypothetical fitness landscape with respect to two traits, which might be parameters in a model.

visualize, the process involves moving around in billions of dimensions, which is near impossible to fathom. But the principle is the same: The goal is to find a fitness peak.

From a given starting spot, an algorithm like this will push the system toward whatever fitness peak is closest, responding only to very local topography, even if there are higher peaks elsewhere. There are multiple ways to avoid getting stuck on a peak that is locally optimal but globally suboptimal. Introducing some changes in random directions instead of only uphill directions—that is, adding some drift—can allow broader exploration of the fitness landscape, possibly leading to a taller hill. We saw this process at work in chapter 6. For real-world evolutionary systems, unusually distant leaps through a landscape can be triggered by combining genes or ideas or components from multiple sources, instead of making small changes to one existing design. These big leaps will most likely end up somewhere worse (lower in the fitness landscape), but occasionally somewhere better. In neural network models, large leaps can be simulated by reinitiating the model with many different random starting points. These tools can ensure that one random starting point didn't lead to a suboptimal solution (i.e., an unusually low peak).

Neural networks are part of a broader family of evolutionary algorithms, which can be used in essentially any situation for which we want to predict some outputs based on some inputs, and for which there are no simple mathematical equations for getting the optimal model parameters. By embracing the idea of outsourcing evolution to the computer, researchers have achieved startling success at an ever-expanding range of tasks, including language translation, medical diagnosis, navigation, designing efficient supply chains, and, of course, human-like conversation. The technology brings with it a range of important moral and ethical questions, which are beyond the scope of this book.

For our purposes, the striking lesson is that research on computer-based intelligence started out as an offshoot of the First Science, working under the assumption that fixed rules or laws could be given as instructions to computers to achieve artificial intelligence. Only by embracing the Second Science—and effectively applying a new Evolutionary Soundboard inside the computer—did AI finally take off. This is another stark reminder of Leslie Orgel's second rule for biology: "Evolution is cleverer than you are."

Artificial Life and Genetic Engineering

All species influence the evolution of other species. Chapter 5 introduced the idea that domestication involves intense selection exerted by one species, such as ants or humans, on another, such as fungi, crop plants, or farm animals. In the early stages of domestication in agriculture, plants that made the biggest fruits or the most seeds, or animals that were the most docile or produced the most eggs or milk, would have been selected as parents of the next generation. The variants were naturally occurring, and selection was partly inadvertent, partly deliberate. Selection of this nature still occurs in modern breeding efforts, but human manipulation of the evolutionary process has gotten a lot more sophisticated.

In addition to tuning the selection dial on the Evolutionary Soundboard, people have been tuning the variation generation dials for several centuries at least. Early efforts began with strategically mating different kinds of plants or animals in order to combine their desirable traits. Imagine one variety of wheat that is resistant to disease but yields relatively little grain, and a second variety that is susceptible to disease but high yielding. If we transfer pollen from one variety to the other, we can expect many combinations of traits in the next generation, depending on the genetic details underlying each trait. We can then select the offspring with the desired combination of strong disease resistance and high yield. If we want a new variety that is most like the original high-yielding one, we can then repeatedly mate the best hybrids with the original high-yielding variety, always retaining only the offspring that are also disease resistant. It could take a while, but ultimately we might produce a variety of wheat that essentially represents the original high yielder, into which a relatively small number of genes have been added to enhance disease resistance. Crop seeds produced this way have been available commercially since the late nineteenth century.

The science of genetics had a massive influence not only on the development of evolutionary theory, but also on the practice of plant and animal breeding. One method for producing new variants is mutagenesis—the application of radiation or chemicals to increase the DNA mutation rate. The odds of producing organisms with more desirable traits this way are small, but with the possibility of screening many offspring for desirable traits, the effort can be worth it. Among thousands of seedlings, there might be one or two that show superior growth, drought resistance, or nutritional value. Thousands of plant varieties have been produced this way, including varieties of rice, corn, soybeans, and ornamental flowers. Using modern techniques in molecular biology, the screening process can be sped up: With knowledge of the genes that underlie particular traits of interest, DNA analysis of seedlings or newborn animals can identify individuals with desirable traits without having to wait for those traits to be expressed (e.g., seed or milk production). This is called marker-assisted selection.

Controlled matings, mutagenesis, and marker-assisted selection all involve naturally occurring variants or a great deal of luck. The process is slow, the chances of success small. As we saw in chapter 4, new variants in evolutionary systems only occasionally result in improvements. The evolution of technology that allows direct manipulation of the genome has drastically increased the speed and chances of success, fundamentally changing the ways in which humans create new biological variants.

In the early 1970s, Herbert Boyer and Stanley Cohen created the first organisms with DNA taken from a different species. In bacterial cells, DNA consists of not only the primary genome (one large chromosome) but also plasmids: much smaller, circular DNA molecules that can replicate and be inherited by daughter cells. Using enzymes that cut DNA in specific places and others that make repairs, Boyer and Cohen took plasmids from the bacterium *Escherichia coli*, inserted genes from different bacterial species or even a frog, and then put these new plasmids back into *E. coli* cells. These transgenic bacteria were able to go on and reproduce, carrying the foreign DNA into future generations.

Despite the success stories from genetic engineering, until recently the methods have been, in some ways, quite crude. One of the most famous and globally important genetically modified organisms (GMOs) is Bt corn, so named because it contains genes taken from the bacterium *Bacillus thuringiensis*. The inserted genes confer resistance to insect attack. They were added to the corn genome using technologies with wonderfully evocative names: nanobullets and a gene gun. Each bullet is a tiny speck of gold or tungsten with a bunch of DNA (the genes you want to insert) on its surface. The bullets are fired at target cells using the gun, penetrating the cell wall and membrane. Once inside the cell, there is some small chance that the new DNA will be incorporated somewhere in the plant's genome, as the plant cells are constantly undergoing DNA replication and repair. So, while on one hand the technology is quite sophisticated (i.e., you couldn't do it at home), it is both literally and figuratively a shotgun approach—you shoot a bunch of DNA at some cells and hope it occasionally sticks without disrupting normal functioning. Most of the time it doesn't work, and even when it does work, it's not entirely clear how and why.

In the past decade, the precision and ease of genetic engineering have taken a major leap forward with the development of CRISPR-Cas technology. The CRISPR-Cas process used by bacteria to incorporate viral DNA into their genomes was introduced in chapter 4. The focus there was on directed mutation and the dynamics of bacteria-virus interactions, which are not the main reasons CRISPR-Cas has received so much attention. The scientific advance that won the Nobel Prize in 2012 was the development of a tool for gene editing. A single RNA molecule can now be easily designed to work with the Cas9 enzyme (there are other Cas enzymes) to cut any piece of DNA at a precise location, where another sequence can be added using the cell's normal DNA repair machinery. In short, shotgun approaches have been replaced with a laser-focused pinpoint method, which now dominates the world of genetic engineering.

It is difficult to overstate the importance of genetic engineering. In the United States, more than 90% of many leading crops, including corn, soybeans, and cotton, are derived from genetically modified seeds. Few genetically modified animals are part of the food system—salmon is one exception—but most farm animals are fed with genetically modified crops. The production of antibiotics has long involved microorganisms that have been genetically modified to produce specific biochemicals, although many bacteria have evolved antibiotic resistance. In response, researchers are now developing benign bacteria to deliver CRISPR-Cas systems specifically targeting pathogenic bacteria that are resistant to the usual set of antibiotics, essentially turning the bacterial defense system back on itself. These bacteria-fighting bacteria are not yet used in human patients, but it might not be too long before they are. Bacteria are also being engineered to do a variety of other jobs, like degrading plastics and sequestering atmospheric CO_2.

In 1996 Dolly the sheep was born—the first cloned mammal, having started as an egg into which the nucleus from an adult sheep cell had been inserted. Cloning has since been applied to pigs, deer, and horses, among others, and some scientists dream of now resurrecting extinct species like the woolly mammoth, whose genome is known from frozen specimens. One approach would be to use CRISPR-Cas to edit, bit by bit, the elephant genome, which already shares 99% of its sequence with woolly mammoths. If that sounds far-fetched, consider that most of what scientists can do today with biotechnology also seemed far-fetched a few short decades ago.

On one hand, it might seem like genetic engineering—as the term suggests—is closer to the First Science, in that engineers decide exactly what they want, and then they build it to precise specifications. Even if this were the case, it would still result in new variants that then enter the dynamics of an evolutionary system. But even the process of engineering isn't quite as precise as it might seem. There is always a phase during which trial-and-error selection is required. For crops, it might involve screening massive numbers of plants for desirable traits. Even with more precise methods, most new organisms fail to survive long enough to do anything useful. For microbes used to produce antibiotics or to fight antibiotic resistant bacteria, the selection phase might involve letting a population go for a time in the environment where they are expected to perform (e.g., inside a mouse). Mutation and selection in that environment essentially fine-tune the population to be able to survive and thrive in the environment where it is expected to perform a specific task. At least so far, even as genetic engineers create biological variants that would otherwise never exist, they have not escaped the need to use evolution itself in their process of doing so.

For millennia, people have been tuning the dials on the Evolutionary Soundboards of a great many species. At first, they probably tuned only the fitness function dial (Dial 5), and probably by just a bit. More recently, they have also dialed up the rate of variation generation (Dial 1), broadened the range of trait values in new variants (Dial 2), brought in genes from other species (Dial 3), and cranked selection way up to keep only the most desired outcomes (Dial 5).

Memetic Engineering

Imagine a world with no marketing or advertising, apart from short, factual descriptions of different products and services. Selection in the economic marketplace would be driven only by the substance and price of each product and how well those matched the needs and desires of consumers. Setting aside price for the moment, you would go to the movies that sound the most entertaining, buy the clothes that seem most comfortable, durable, and practical, and consume the beverages that were the tastiest and most nutritious. Selection would thus drive the evolution of more entertaining movies, more comfortable and durable clothes, and tastier beverages. The problem is that different options might be effectively indistinguishable on these axes.

Consider the case of the two dominant American soft drinks of the twentieth century: Coke and Pepsi, both caffeinated, carbonated, sugary colas. How to choose? There are some small measurable differences between the two beverages. In a 355 milliliter can, Pepsi has more caffeine (37.5 milligrams) and sugar (41 grams) than Coke (34 milligrams and 39 grams, respectively), while Coke has greater carbonation. However, blind taste tests are typically inconclusive, suggesting that these differences are undetectable by the human palate, or at least insufficient to drive consistent preferences. If there is any small average preference, it is actually in favor of the less popular brand, Pepsi.

From an evolutionary point of view, we can think of "blind" satisfaction as a component of economic fitness—all else equal, greater consumer satisfaction from a given product will lead to an increase in its prevalence in the market. The fitness functions describing how satisfaction relates to beverage characteristics (traits) seem quite likely to be hump-shaped in most cases (fig. 7.4, top panel): Too little or too much sugar, caffeine, or carbonation is unsatisfying, with greatest satisfaction achieved at some intermediate level. However, given that the differences between brands are very small, it also seems likely that Coke and Pepsi fall on a flat portion of the fitness functions. That is, the differences are too small to matter, which can explain the observation that there is no overall blind preference.

As a teenager, my group of friends could argue at length about the superiority or inferiority of Coke versus Pepsi, McDonald's versus Burger King hamburgers, Titan versus Sher-Wood hockey sticks, or Nike versus Reebok sneakers. We were convinced that the differences were big and real, but a close look at all these brands and countless others shows them to be almost indistinguishable with respect to most characteristics one can think of—taste, durability, comfort, etc. In other

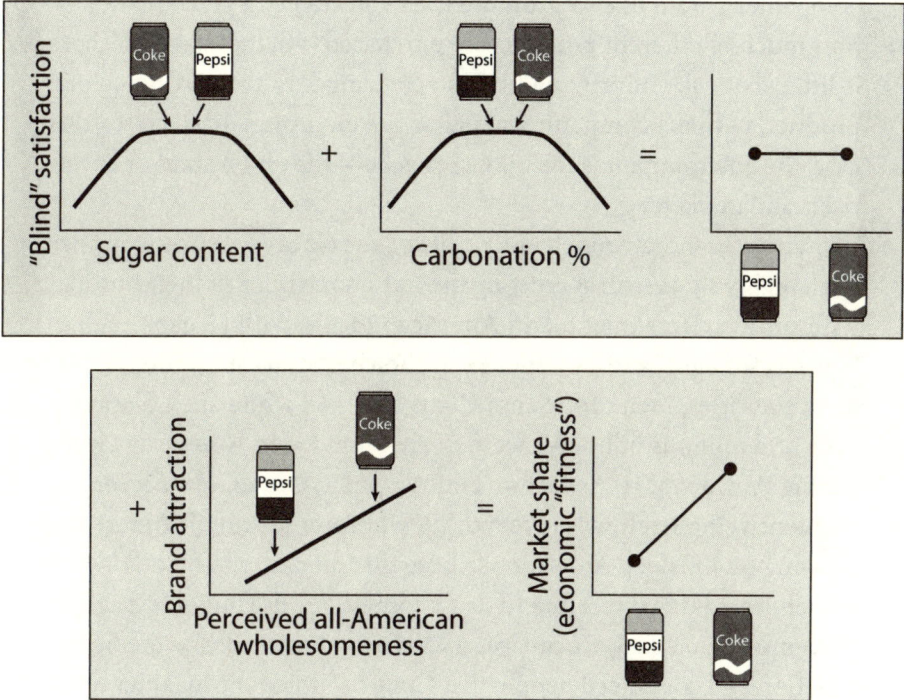

FIGURE 7.4. Hypothetical consumer satisfaction as a function of the traits of Coke and Pepsi (*top*), and the influence of advertising on market share (*bottom*).

words, the economic world is replete with products populating flat portions of fitness functions, at least according to most of their functional characteristics.

Enter marketing and advertising, which are, at their core, attempts to guide the course of cultural evolution. Culture can be thought of as the sum total of what people think, believe, make, and do, and marketers and advertisers essentially attempt to exert control over what *other* people think, believe, make, or do. This very book is an attempt to influence the way people think about science, and the title of this section—"memetic engineering"—was an attempt to grab your attention and pique interest. (I don't personally find the concept of the meme—as a small unit of culture—especially useful, but I found the expression "memetic engineering" irresistible as a bridge from the previous section; I vow to not write it again in the rest of the book.) In the economic realm, companies would like to promote the evolution of a given market—how much of different products are purchased—in their favor. If there is little scope for improving fitness via changes to the substance of a product, perhaps companies can alter the environment itself. In this case, the environment is the human psyche—the playground of advertisers and marketers.

Starting in the late nineteenth century, the Coca-Cola Company has been highly successful at creating mental associations of their product with purity, refreshment, and American ideals, with slogans such as "Pure as sunlight" and "America's real choice." Ads have featured countless celebrities, including Santa Claus (red and white are Coke's colors), and millions of bottles were freely provided to American troops during World War II. As author Tim Wu put it, Coca-Cola "succeeded by identifying itself with everything wholesome and all-American, drawing on the deep American self-regard and desire to belong—and somehow making it feel that to drink something else might be vaguely treasonous." In effect, Coca-Cola added new dimensions to the fitness function and convinced people that Coke occupied the maximally all-American and wholesome portion of that space. A flat fitness surface based purely on content was transformed into a steep one based on imagined associations (fig. 7.4). For most of the past century, Coke has dominated over Pepsi.

In their attempts to create new axes of variation to distinguish a brand, advertisers frequently co-opt components of selection related to evolved biases in our choices of whom to copy habits and behaviors from. In deciding whom to emulate, we are prone to at least two major kinds of bias, as discussed in chapter 5. The first is "prestige bias": We are far more likely to emulate someone who is famous than someone who is relatively unknown, and advertisers hope that prestige bias will function even when the product is unrelated to the source of prestige. While there might be some rational basis for copying Michael Jordan's choice of basketball shoe, there is no reason to suspect that I will derive greater enjoyment from the brand of coffee that George Clooney is hawking than from any other brand. Celebrity endorsements are effectively an economic application of prestige bias in cultural evolution, albeit with equivocal evidence with respect to how well they actually work.

The second important bias is based on the frequency or prevalence of different types. Group belonging and identity are of profound importance in human psychology, whether based on religion, geography, politics, common values, or just supporting the same sports team. In many such cases, we choose to follow the crowd. In chapter 5 we encountered positive frequency-dependent selection, which involved a runaway process by which crossing a threshold of prevalence provided a systemic advantage to the more common type, leading to dominance of a particular choice of word, software, or hit song. This dynamic presents an opportunity for advertisers—such as those trying to get you to buy more Coke.

If advertisers are faced with undetectable differences in the substance of different products, they can nonetheless hope to associate their product with a group identity (e.g., being American) that exploits or even creates a positive frequency-dependence. Not only can they try to tip the balance in favor of Coke (or any other brand), as illustrated in figure 7.4, but they can hope to sit back and watch "natural" dynamics take over once an initial degree of success has been achieved. Even if a superior product comes along, positive frequency-dependent selection can make it difficult to penetrate the market.

For an upstart or underdog product, advertisers can also, at least initially, try to harness the opposite phenomenon: negative frequency dependence. Certain types of consumer might prefer to buy whichever type is *not* dominant, thus presenting another quirk of the human psyche for advertisers to exploit. For a time during the 1960s counterculture movement, Pepsi had some success with this strategy, proposing "the Pepsi generation." However, once counterculture inched into the mainstream, Coke proceeded to beat Pepsi at its own game, with its unforgettable song, "I'd like to teach the world to sing / In perfect harmony / I'd like to buy the world a Coke / And keep it company."

Just like plant breeders and genetic engineers, marketers and advertisers are effectively in the business of tuning the dials of an Evolutionary Soundboard. When faced with a fitness function dial (Dial 5) set close to zero (similar fitness of different variants), they attempt to tune it in their favor, either by creating new trait axes or by exploiting transmission biases. They also pay close attention to Dial 6 (the frequency dependence of selection) to know whether there is an opportunity for runaway success (positive) or a niche market (negative). While we tend to associate advertising with making money, the same basic techniques of persuasion are used toward a variety of ends, most notably in politics and religion. In these contexts, efforts to direct the course of cultural evolution are often labeled as propaganda, and they often involve restricting the flow of information—that is, reducing the number of influences on people's behavior (Dial 4).

This section has been about people influencing other people. But humans are not alone in doing these things, even if our methods involve an unprecedented degree of sophistication and complexity. Many butterflies, which would otherwise be perfectly palatable to predators, have evolved the same color pattern as poisonous species, thus deceiving potential predators. When ravens or chimpanzees are under attack, they vocalize in ways that depend on their audience—more volume and drama if likely helpers are nearby. In many animal species, the biggest, strongest, and cleverest individuals frequently exert power over others, for example by excluding the less powerful from matings. In cuttlefish (cousins of octopus and squid), subordinate males can alter their coloration and retract one of their four arms in order to resemble females (who have three arms), and thereby sneak past a dominant male to mate with an otherwise protected female. In short, persuasion, deception, and power dynamics are involved in many evolutionary systems, not just human cultural evolution.

For many animals, there will be a selective advantage to *resisting* persuasion or deception. In other words, adaptive evolution of one of your traits (ability to detect deceptive communication) can oppose outside attempts to manipulate the cultural evolution of a different one of your traits (e.g., consumption patterns). Deceptive communication in animals is thought to persist only when relatively rare. If, for example, a nonpoisonous butterfly is abundant compared the poisonous version it mimics, either the cost to a predator of making mistakes will be low (may as well just eat everything), or there will be strong selection favoring discrimination. In the human world of advertising, people can quickly become immune to attention-grabbing tactics, and their tolerance of being bombarded with ads can be pushed only so far. As a result, the history of advertising has shown repeated cycles of new, innovative approaches to capturing people's attention (e.g., excessive ads on television), ultimately leading to revolts from the target audience, and massive reconfigurations of the advertising industry itself.

Repeated reconfigurations of the advertising industry show us that evolutionary change is not always slow and gradual. In this situation and many others, sometimes tipping points are crossed, leading to complete reorganization of the system. This is the topic of the next chapter.

8

Tipping Points and Major Transitions

Two of your friends want to borrow money from you. Kai has a stable job and is generally reliable, so you keep the interest rate low for Kai. Avery is in a more precarious financial situation, so you charge Avery a higher interest rate, given the greater risk of not being paid back. Unbeknownst to you, however, Avery has already borrowed money from Kai. In banking terms, Kai has financial "exposure" to Avery that you didn't know about, so your overall risk in lending money to your two friends is greater than you thought. In the end, Avery defaults on both loans, and in an instant, everyone is screwed. If all relevant information had been shared with everyone else, the situation might have been averted. You would have charged higher interest rates, which might have reduced the amounts of each loan, or maybe even prevented one or both transactions to begin with. In evolutionary terms, selection for maximal profit under incomplete knowledge led to risky lending behavior, making the whole system susceptible to collapse. And in this simple scenario, we can glean the key message of this chapter: Evolutionary systems can sometimes be prone to sudden, profound changes.

Something like our little lending triangle—except far more complex—happened in the lead-up to the global financial crisis of 2008.[1] In any economy, there are many interacting evolutionary systems. In general, selection favors activities that increase the short-term economic success of different entities, be they countries, banks, or investors. In the decades leading up to 2008, variation generation and selection drove evolutionary changes in government regulations, in the lending practices of banks, and in the structure of different investment products. First, relaxed government regulation of banks in the United States, combined with low interest rates, selected for altered lending practices, with mortgages increasingly offered to people who would otherwise be considered too risky (like Avery). Demand for homes rose, as did prices, and the resulting boom in the housing market selected for an increased prevalence of investment products based on mortgages. Specifically, thousands of mortgages were packaged and sold to investors as "mortgage-backed securities"; in essence, investors paid to take on both the risk and benefit that might accrue to the mortgage providers. As long as interest rates stayed low and home prices rose, everyone would benefit. But the system overall had evolved a major fragility.

The less-than-perfect transparency of financial institutions, and the bewildering complexity of many investment instruments, led to systematic underestimates of risk. So, when interest rates rose and the holders of risky mortgages were forced to sell their homes, home prices fell, and financial losses spread rapidly among many densely connected institutions. Institutions thought to be too big to fail failed. After the fact, scientists searched for potential warning signals leading up to the crisis. One study discovered an increased prevalence of lending relationships among banks just like the one between you, Kai, and Avery—that is, relationships in which bank A lends to bank B (and/or bank C) without knowing about a lending relationship between B and C. Why exactly these shifts in interbank relationships came about is not clear, but their prevalence was likely one element in the complex set of causes or symptoms of the compromised resilience of the whole system. Selection pushed an evolutionary system up to, and then beyond, a tipping point.

When an evolutionary system undergoes a massive change that is difficult to reverse, scientists call the change a *critical transition*. Whether we're talking about an ecosystem, a set of cultural norms, or an economy, certain configurations or states of the system are relatively stable, in the sense that they tend to return to their initial configuration following disturbances, such as a rise in interest rates or climate warming. But in some cases, if the system is pushed beyond a tipping point, it can rapidly shift to a completely different configuration that is also persistent. This means that even if the driving factor is reversed (e.g., interest rates are lowered, the climate cools) the system will be stuck in its new state. These situations are of great concern in many domains, given the great difficulty in predicting where the tipping points lie, and the potentially massive efforts required to reverse them.

In this chapter we'll focus on major, critical transitions and tipping points. We have already encountered some situations of this nature, specifically under positive frequency-dependent selection (chapter 5), and in the efforts of advertisers to push cultural choices beyond tipping points to achieve market dominance (chapter 7). Here we will revisit this topic from a somewhat different perspective and then dive into case studies from the natural and cultural worlds where complex combinations of negative and positive feedbacks make evolutionary systems susceptible to critical transitions.

Balls Roll Downhill

My main research site is in Mont Mégantic National Park, in southern Québec, Canada. As we enter the park at low elevation, where conditions are relatively warm (for Canada), we are in forest dominated by broadleaved trees, mostly sugar maple, but also some American beech and yellow birch. The cold mountain top is dominated by conifers, mostly balsam fir, but also red spruce. The abruptness of the transition from one forest type to the other (see fig. 5.5), in the absence of any abrupt change in soil or climatic conditions (both of which change gradually with elevation), suggests the possibility of two stable states at the cool mid-elevations: dominance by broadleaved trees or by conifers, but rarely an even mix. This would be the expected outcome of positive frequency-dependent selection, by which a given type of tree gains a fitness advantage when it starts out more abundant than the other.

An evolutionary explanation for how forest composition changes on this mountainside is show in figure 8.1. At all elevations, the dial for the

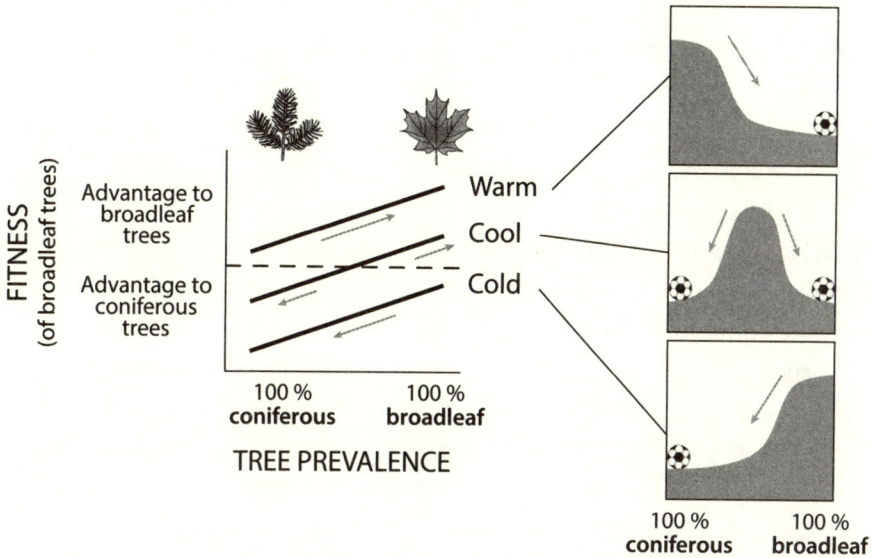

FIGURE 8.1. Interactions between positive frequency dependent selection (the fact that the lines in the graph on the left have positive slopes) and directional selection (the fact that each line has a different vertical position, depending on the environment). Arrows indicate the expected direction of change from any starting point.

frequency-dependence of selection (Dial 6) is tuned to the same setting—positive—reflecting a relationship between the fitness of broadleaf trees and their prevalence that always has the same positive slope (fig. 8.1, left). For a given individual tree, the more its neighbors are of the same type, the greater its fitness. The fitness function dial (Dial 5), however, is tuned to *different* settings at each elevation: The position of the sloped lines moves downward as we go from warm to cold places. At the warm, low elevations, broadleaf trees enjoy a selective advantage, shown by the warm line being in the upper part of the graph. This general advantage is strong enough to overcome any tendency for reduced fitness at low prevalence: Broadleaf trees always "win" at low elevation, even if they start out relatively rare. Same thing for the conifers at the cold, high elevations, where a strong *dis*advantage for broadleaf trees implies a strong advantage for conifers.

At mid-elevations, things get more interesting. The sloped line in figure 8.1 crosses the threshold (the dashed line) between coniferous trees having an advantage and broadleaf trees having an advantage. As such, predicting which species wins will depend on their starting frequencies. Specifically, whichever starts out at higher prevalence should come to dominate a given spot in the forest. Nowhere do we expect an even mix of the two types of trees.

Another way of visualizing this dynamic is to think of a ball on a hilly surface, with the environment defining the shape of the surface (fig. 8.1, right). Under both warm and cold conditions, there is only one low point, so the ball always tends to end up there, regardless of where it starts. Under warm conditions, we get dominance by broadleaf trees; under cold conditions, we get conifers. Now, what happens if we perturb the forest by cutting down trees and planting the "wrong" kind: conifers in warm places, broadleaf trees in cold ones? In warm places, the perturbation moves the ball to the left, while in cold places the ball moves to the right. But in both cases the ball is being pushed uphill, after which we expect it to roll back down, returning the system to its initial state. In contrast, at the cool mid-elevations, there are two low points, and so moving the ball right or left might push it over the top of the hill. In other words, at mid-elevations we cannot always count on the system returning to its initial state after a perturbation. If we cut down all the trees in a conifer-dominated area and replant with maples, we do *not* expect a return of the conifers—the area should stay dominated by maple.

To generalize beyond forest types, the x-axis in these ball-and-surface diagrams can represent any state of any evolutionary system: the dominance of corals in a reef, the lending behavior of banks, or the prevalence of a particular technology. The general situation of interest is when there are multiple stable states, represented by more than one low point on the surface (there could be more than two, but two low points suffice to make the important points here). Although the forest example and others mentioned in chapter 5 emphasized positive feedbacks, scenarios involving critical transitions and tipping points actually involve a mixture of both negative and positive feedbacks.

Consider a situation with two stable states, A and B. When a system is in state A (fig. 8.2, top), negative feedbacks are represented by the slopes on either side of that point, returning the ball to state A if it gets

Examples	State A	State B
Reefs:	Coral —— Algae	
Lending habits:	Permissive — Strict	
Keyboards:	QWERTY — Fġğıod	

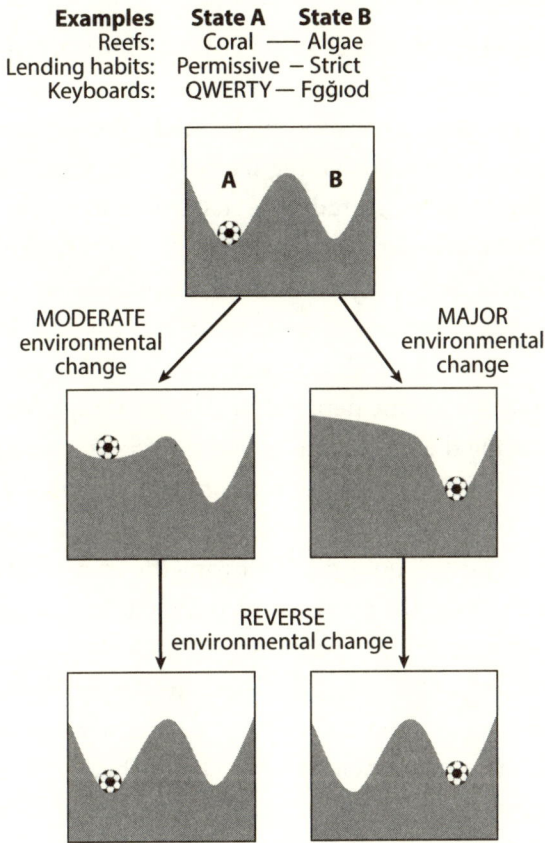

FIGURE 8.2. Positive feedbacks (the tendency to roll away from a hilltop), negative feedbacks (the tendency to stay in a valley bottom), and how a major environmental change can cause the system to cross a tipping point.

pushed a bit left or right—that is, if there is some perturbation that isn't too large. Perturbations might include a storm that kills some corals or an upstart bank offering an unusually low interest rate. If a system finds itself on the peak of the hill between states A and B, in principle it could stay there indefinitely, in the sense that the surface at precisely that point has no slope, and we could imagine balancing the ball there, just so. But the smallest of perturbations would send it running downhill, away from this unstable equilibrium to either A or B (stable equilibrium points), depending on the direction of the perturbation; this is what is meant by positive feedbacks.

Just as the shape of the surface varied among different climatic conditions in the forest example, a change in the environment over time will also alter the strength and potentially the nature of the feedbacks. Environmental change might involve increased fishing pressure in a coral reef or a tightening of financial regulation by the government. If the change is of moderate magnitude, the stability of state A might remain intact, but with a shallower depression (fig. 8.2, middle left). This means that the resilience of the system has declined, in the sense that it takes less of a push to get over the hill in the middle, and it takes longer to roll back to state A when indeed it does roll back. With a much stronger environmental change, the depression at state A might be eliminated altogether, causing the ball to roll into state B (fig. 8.2, middle right).

The key lesson of this story comes from what happens when we reverse the environmental change, and thus when the surface returns to its initial shape (fig. 8.2, bottom). Following the reversal of a moderate environmental change, the system remains in state A, as resilient as ever. In contrast, after reversing a major environmental change, the system stays in state B, and negative feedbacks can keep it there indefinitely. To get the system to return to state A, reversing the environmental change isn't enough; the environment needs to be pushed further in the opposite direction than where it started. For example, if the initial environmental change was climate warming, we might require substantial climate cooling—well below the initial temperature—to raise the right side of the surface enough to cause the ball to roll back to state A. Scientists use the word *hysteresis* to refer to this kind of dynamic, in which the state of a system depends not only on the environment, but also on the history of the system itself.

Although it might seem simplistic to think of the states of ecosystems or economies as determined by a ball rolling on a surface, this analogy actually captures quite accurately the dynamics of the sophisticated models built to characterize complex evolutionary systems. For example, the shallower slopes adjacent to state A after moderate environmental change (fig. 8.2, middle left) suggest a potential early warning signal of a critical transition: a marked decrease in the rate of system response to small perturbations. Balls roll down shallow slopes more slowly than they roll down steep slopes. This "critical slowing down" is exactly what is seen in the lead-up to critical transitions both in mathematical models and in time-series data for real-world ecosystems and societies. Despite a great variety of specific mechanisms at play, many different biological and social systems show clear signs of being prone to hysteresis.

Ecosystems

Coral reefs are dominated by, well, corals. The elaborate three-dimensional shapes we see in a coral reef are not individual organisms, but large colonies of little invertebrate animals called polyps, each of which has secreted an exoskeleton around itself. It's the collective result of thousands of polyps stuck together that produce the large, sometimes colorful structures that look like brains, fingers, shelves, leaves, or shrubs. Also growing in these reefs are large, leafy macroalgae, although invertebrates like sea urchins along with some kinds of fish graze the macroalgae, thereby keeping them in check. The complex structure of corals contributes to keeping the macroalgae in check by providing habitat for fish and places for urchins to hide from predators. Taken together, this set of interactions creates negative feedbacks that can keep reefs in a coral-dominated state. If a mild disease outbreak kills some corals, they should grow back.

Larger disturbances can trigger a critical transition. Heavy fishing and urchin disease outbreaks can overcome the negative feedbacks and tip the competitive balance in favor of the algae. Much of the energy acquisition in corals comes from photosynthesis done by symbiotic microalgae living within the tissues of the polyps, and so they are dependent on light. When the large macroalgae are released from herbivory pressure that would normally come from fish and urchins, we can expect a decline of corals via shading, which in turn eliminates the important structural features of the habitat for fish and urchins. So even if there is a reduction of fishing pressure and the disease peters out, the system can be stuck in a new macroalgae-dominated state.

On the Evolutionary Soundboard for the coral reef, the fitness function dial has been moved temporarily to favor algae, which, according to the frequency dependence dial, permits a switch to algae dominance. Returning the fitness function dial to its initial position is not sufficient to reverse the change—one must move the dial further in the opposite direction to achieve that. It takes a much bigger push to get out of the new state than it took to get into it.

Another well-studied example of hysteresis comes from shallow lakes of the temperate zone, where the growth of aquatic plants rooted in the lake bottom is limited by the availability of nutrients. Photosynthesis in these lakes is also done by the tiny phytoplankton floating freely in the water. Phytoplankton are eaten by the slightly larger zooplankton, and so healthy zooplankton populations keep the phytoplankton in check. Zooplankton are eaten by fish, but with a sufficient abundance of rooted plants, the zooplankton can still thrive because the plants give them somewhere to hide from the fish. Just like in the coral reef, there is a set of negative feedbacks maintaining ponds in a state with many rooted plants, lots of zooplankton, and fairly clear water.

When agricultural runoff increases nutrient input into shallow lakes, the rooted plants grow to the point of becoming a nuisance, so people sometimes yank them out. Without the plants, the zooplankton have nowhere to hide, and the fish drive down the zooplankton populations, thus allowing the phytoplankton to grow out of control. With no roots holding together the mud on the lake bottom, the water column gets quite cloudy. With thriving phytoplankton and muddy water blocking light from the bottom, seeds or seedlings of the larger plants are unable to reestablish themselves. At this point, even if nutrient input is reduced, the system is stuck in its new phytoplankton-dominated, muddy state. The dials on the Evolutionary Soundboard have been tuned just as they were for the coral reef example.

Biologists who study forests, coral reefs, or lakes have the luxury of being able to replicate observations across many independent but closely similar places, often with the possibility of manipulating things experimentally—cutting or planting trees, removing fish or providing refuges for them, adding nutrients, etc. This allows them to work out some of the mechanisms underlying state changes, and to build predictive models of how the system works. In studies of human populations, the scope for replication and especially experimental manipulation (except in highly contrived lab situations) is much reduced, which means that we often have somewhat less confidence in piecing together underlying mechanisms. That said, there are plenty of compelling examples to suggest that socioeconomic systems are frequently subject to critical transitions and tipping points as well.

Cultural Norms

Everyone at my high school agreed: Ms. MacBride had eyes in the back of her head. Ms. MacBride taught typing, and no matter where she was in the classroom or what she was doing, she would somehow know if you dared glance down at your fingers on the keyboard. "VELLEND! EYES UP!" Needless to say, she was not our favorite teacher at the time, although I can now think of few other specific skills learned in high school that have been as useful in my career as ten-finger, eyes-on-the-page typing. Thank you, Ms. MacBride.

One of the reasons it is so tempting to look down at one's fingers while typing is that the arrangement of the letters on the keyboard makes no obvious sense. Starting at the top left, the order of letters begins with QWERTY, from which the "QWERTY keyboard" takes its name. The history of how this particular layout came to dominate the market has been the subject of surprisingly contentious academic debate, but one thing seems clear enough: The selective pressures on keyboard design that applied when QWERTY was introduced in the late nineteenth century are no longer relevant in the twenty-first century. Yet we're still stuck with QWERTY.

On early typewriters, pushing down on a letter—let's say B—caused a small metal rod called a typebar to swing up from its hanging position toward a sheet of paper. Embossed on the tip of the rod was an inverted B, and when it struck an ink ribbon, the letter was typed onto the paper. Some forty-odd typebars—letters, numbers, punctuation marks, and a few symbols—were arranged in a tight circle called a typebasket, so that each one would slap the paper in the exact same spot when lifted toward the middle of the circle. If two adjacent typebars came up in quick succession, they could easily get jammed together, annoying and slowing down the typist considerably (this was still true for later models with a semicircular typebasket, like the one in my childhood home). The designer of the QWERTY keyboard, Christopher Latham Sholes, never revealed all his inner thoughts, but there is little doubt that one important design criterion was minimizing the frequency of jamming. He seems to have ensured that pairs of adjacent letters or symbols in the typebasket were among those that tended to occur infrequently as a pair in written English. (Note that the order of typebars in the typebasket was not identical to the order of their corresponding symbols on the keyboard; for example, in the typebasket E and R were separated by S, which was one row up from these two letters on the keyboard.)

Eventually, the jamming-prevention criterion became irrelevant. This was true even on the electric typewriter I used in Ms. MacBride's class in the 1980s, in which the letters and symbols were embossed on a ball that turned to point the appropriate letter at the piece of paper. It hardly requires mentioning that jamming is also irrelevant on the keyboards or keypads now used for virtually all typing with computers and phones. Not surprisingly, alternative keyboard layouts that increase typing efficiency have been invented, and they can easily be implemented. Yet QWERTY (or minor variations on QWERTY) are still used across most parts of the world where languages based on the Roman alphabet are written. Even in Turkey, where the government mandated a new layout called Fgğıod in the 1950s to increase efficiency (several world records for typing speed were set), the international spread of computers brought QWERTY back to dominance.

The case of QWERTY is one example of what economists describe as "path dependence" and "lock-in" of standards. These terms imply the presence of frequency-dependent selection, both positive and negative. Whether the standard in question involves systems of measurement (metric versus imperial), the shapes of plugs and outlets, audio and video recording media (records, CDs, or digital files), or file transfer protocols, there are benefits to having a common standard, and costs to switching. This creates negative feedbacks, which tend to keep the system in its current state—in this case, using the old standard.

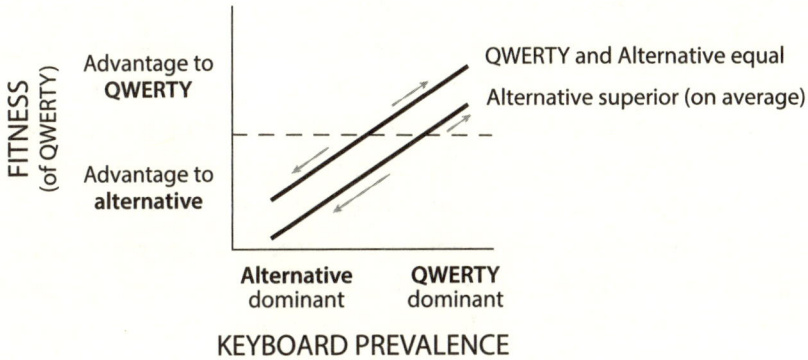

FIGURE 8.3. Hypothetical positive frequency dependent selection that maintains the dominance of QWERTY keyboards, despite equally good (*top line*) or even better (*bottom line*) alternatives.

Just inventing a superior alternative might not be enough to get people to break out of old habits. In figure 8.3, the upper relationship represents a situation in which QWERTY and some alternative design are equivalently good. Superiority of the alternative (on average) is represented by lowering the position of the relationship shown by the diagonal line. If the new relationship still crosses the horizontal zero line, QWERTY can still maintain its dominance, but the threshold for change has moved closer to the QWERTY-dominant state on the right. So, if enough people made the switch (roughly 20% according to the graph), a tipping point would be crossed, after which positive feedbacks would kick in, prompting a rapid transition to the new standard. In some cases, such as audio recording media, forces have been strong enough to cause repeated waves of rapid changes in standards: vinyl records, cassettes, CDs, etc. In the case of QWERTY, and others where we might feel stuck, perhaps it is only a matter of time before a tipping point is crossed. Or perhaps the advantages offered by QWERTY alternatives are just too small to ever overcome the inertia that keeps things as they are.

The potential for critical transitions in cultural norms appears to be widespread. It is not difficult to think of examples from everyday life in which there is pressure to conform: the layout of cutlery in table setting, fashion choices, attitudes toward homosexuality, smoking in public places, or food aversions and perceived delicacies. For some of these, such as fashion, the alternatives are of little functional consequence. In other cases, adhering to a norm can involve steep, even gruesome costs. For hundreds of years, up to the end of the nineteenth century, many women in China had their feet broken, bent, and bound, to make them permanently small, bowed, and pointy, compromising their ability to walk. As a sign of social status and imposed fidelity (the woman could hardly leave the house), foot-binding enhanced marriage prospects, so the practice was difficult for individual families to abandon. Repeated efforts at prohibition failed. It was only after concerted grassroots efforts that the practice finally fell out of favor, early in the twentieth century. During roughly the same period of time that foot-binding was practiced in China, dueling to preserve one's honor was common practice among upper-class European men—accompanied by the risk of death or serious injury. Female genital mutilation occurs in some parts of the world to this day. In all these cases, strong social forces can allow cultural norms to endure for centuries, despite massive costs.

While some social norms can become formalized in the form of laws or regulations, most often they appear to evolve spontaneously via repeated social interactions that involve pressure to conform. When there are rewards for conforming (e.g., enhanced marriage prospects) and sanctions for deviating (e.g., social ostracization), cultural norms arise spontaneously. Using some clever experiments, researchers have recreated this process in the lab. Sociologist Damon Centola and colleagues created groups of online strangers given the task of assigning names to photographs of human faces. Members of a group met two at a time, with randomly chosen partners, and each member of a pair was rewarded if they had assigned (in advance) the same name to a photo. As these pairwise interactions proceeded, the only information available to participants was their record of past interactions—what names they and their previous partners had assigned to which photos. Fairly quickly, such groups converged on cultural norms (called "conventions" by the researchers) in terms of which names were given to different photos. Different groups converged on different norms, which was expected given that rewards accrued for agreeing on *any* name, not a specific name. For example, one group might have converged on the name

Alvin for a given photo, while a second group converged on Simon for the same photo.

In a second round of experiments, it was also possible to push an existing norm across a tipping point to create a new and different naming convention. Having established an initial norm, some groups were infiltrated by a small set of "confederates" of the researchers, coordinating to push for a new norm. If these confederates made up less than 25% of the people in the group, they generally failed to effect change. But beyond 25%, they could tip the balance toward a new norm—prompting a switch from Simon to, let's say, Theodore. Although experiments like these create situations that are far simpler (and more trivial) than those involving norms like foot-binding or attitudes toward sexual diversity, they do provide a proof of concept that social pressures can create positive frequency-dependent selection, and thus the potential for tipping points and critical transitions.

On the Evolutionary Soundboard for all these cultural examples, new variants were introduced (Dial 1) with new traits (Dial 2) and new average fitness values (Dial 5). As we can see in figure 8.3, however, whether altered average fitness actually changes the prevalence of different types will depend on the strength of positive frequency-dependent selection (Dial 6). For evolutionary change to happen, sometimes a major perturbation or unusually strong selection is required to push the system beyond a tipping point or to eliminate that tipping point altogether.

Major Transitions in Evolution over the Long Term

Most of the examples presented so far have involved systems that are stabilized by negative feedbacks, with the potential for flips between different states if perturbations trigger positive feedbacks. These system flips are often called critical transitions, and they have been observed in a variety of biological and cultural systems, typically during time spans no longer than one or a few human generations. Expanding the time frame to cover the history of life on Earth, biologists have also recognized what they call "major" evolutionary transitions. As the name implies, these are unusually large evolutionary changes—something more profound than bird beaks getting larger or legs becoming fins, and less likely to be reversed.

There are no universally agreed on criteria to define major evolutionary transitions, but the clearest cases all involve one distinguishing feature: Previously distinct entities linked their fates into a larger whole, thus creating an entirely new type of evolutionary entity. The best examples are the evolution of life itself (prokaryotic cells), eukaryotic cells, and multicellular organisms. When self-sustaining biochemical reactions became bound within lipid membranes, forming cells capable of replication, life as we know it had begun, and the fates of the biomolecules in a cell became intimately linked. The earliest kinds of cells—prokaryotes—have been around for about 3.5 billion years. Eukaryotes were born more than 2.5 billion years ago, when one prokaryote was ingested by another, but instead of one cell digesting the other, the two entered into a symbiotic relationship. Subsequent evolution has produced complete mutual dependence—neither the larger cell nor its symbiont (mitochondria) could now survive without the other. More than 1.5 billion years ago, evolution gave rise to multicellular eukaryotes, which now include humans. Our bodies are made up of trillions of cells, coordinating their actions in marvelously complex ways to ensure (most of the time) the efficient functioning of the multicellular organisms they comprise. For the most part, the individual cells of our body can advance their evolutionary fitness only by contributing to the success of the whole organism.

Most biologists would have to agree that the formation of a new and distinct type of evolutionary entity counts as a major transition, especially given the spectacular subsequent diversification and success of the descendants in each case. These examples of major transitions are distinguished from critical transitions in leaving little chance for possible reversal once a certain level of complexity and interdependence of parts has evolved.[2] For example, it is highly unlikely that a lineage of eukaryotic cells would disaggregate into free-living prokaryotes, and while multicellularity has evolved from unicellular ancestors more than twenty times, the reverse is quite rare (albeit not impossible).

What major and critical transitions have in common is the interplay of positive and negative interactions or feedbacks, albeit in somewhat different forms. In their free-living unicellular form, prokaryotes or eukaryotes might compete for the same resources or eat one another— ecological relationships that are unsustainable if those cells are to get together and form a functional whole. So, when eukaryotes and mul- ticellularity evolved, there was a transition from largely antagonistic (negative) interactions to highly cooperative (positive) ones. But once unicellular organisms had joined forces, making either larger and more complex cells (eukaryotes) or multicellular creatures, their success would once again be determined by their ability to compete with (or eat) other organisms. When competition transitions to cooperation, the competitive ability of the cooperative alliance is enhanced. And when competition *within* the cooperative alliance does happen—e.g., the proliferation of cancer cells—it can be to the severe detriment of the whole.

The cooperation-competition interplay has also been front and cen- ter in another proposed major transition: the evolution of eusociality, or "true" sociality. Eusocial animals (mostly ants, bees, and wasps) form colonies in which many sterile individuals work in service of the whole colony, whose genetic fitness runs through a relatively small number of reproducers. Eusocial colonies provide one example in the broader category of "societies." The evolution of societies with language—that is, human societies—has been proposed as one of the major transitions in evolution. It's debatable whether humans or their societies represent a qualitatively new kind of evolutionary entity (compared to other animal societies), but regardless, it's clear that the cooperation-competition interplay has also been a key factor under- lying the evolution of human culture.

If you think carefully about how you meet your daily needs and de- sires, there is scarcely a single thing you can do without the cooperation of a sizable network of other people. I am typing on a computer as- sembled by people who are strangers to me, using components pro- duced by yet other strangers, who procured raw materials from still other people. The computer was delivered to me by people driving trucks on roads built with money provided by taxpayers. The food I eat is grown almost entirely by other people, and even the handful of tomatoes and herbs I grow myself required seeds and stakes provided by other people. If I get sick, a team of nurses, doctors, and technicians will help diagnose and treat my illness. Regardless of whether people's actions are motivated by self-interest or altruism, almost everything I do depends on those actions and so can be thought of as cooperation in a broad sense. Left to my own devices in a world without other people, I like to think I could eke out an existence and survive, but I'm probably wrong about that.

For a group of people to succeed—whether in terms of economic, military, or social indicators—direct or indirect cooperation is required. This applies to groups of every size, from single families to entire empires. But how do we define success in such cases? In almost all scenarios, success is a relative concept: Success is doing better than some other group. As conflicts between groups over resources are inevitable, it is thought that intergroup competition has been a major force of selection favoring the cultural evolution of cooperation within groups of people. In turn, enhancing the scale and depth of cooperation within groups (e.g., building a large military) can magnify the intensity of conflicts between groups. Cooperation within groups and competition between groups appear to go hand in hand (fig. 8.4). As evolutionary biologist David Sloan Wilson put it, "It is an inescapable fact that our admirable ability to cooperate within groups is due very largely to violent competition among groups."

The cooperation-competition dynamic relates to the topic of major or critical transitions in two ways. First, intragroup cooperation is intimately linked with cultural norms, which, as we have seen, are often subject to potential tipping points and critical transitions. Intergroup cultural selection favors cooperation within groups, and cultural norms can act as the glue that holds cultural groups together, although those norms can take a great variety of forms. One of the most widespread manifestations of formalized cultural norms, binding people together in cooperative relationships, is religion, which has been blamed for

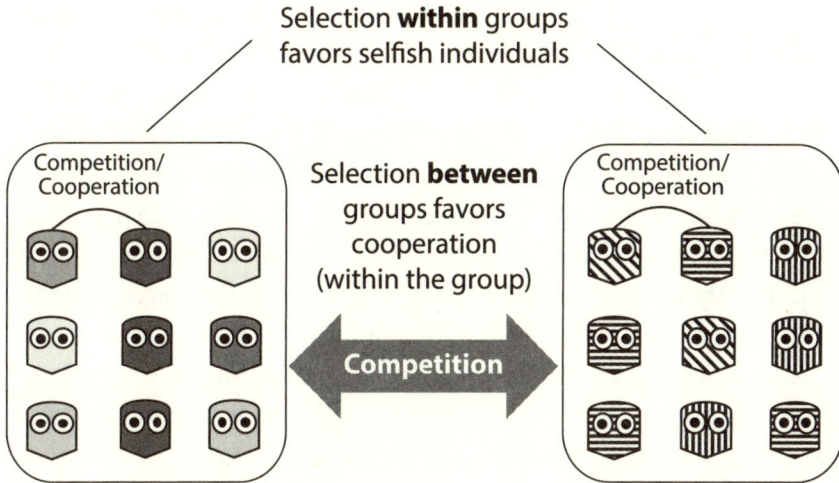

FIGURE 8.4. Within a group, selection favors selfish individuals, who can procure more resources. But if multiple groups are in conflict, success of groups—and so the individuals within them—will depend on cooperation, which is therefore favored by selection.

many wars, although it has also been argued that religion might be as much a *consequence* of war as a cause. Conflict favors the group with greater cooperative cohesion.

The second link comes from research on the causes of major historical events and cycles, such as the dramatic rise and fall of empires. Biologist-turned-historian Peter Turchin has pioneered the application of quantitative approaches to understanding such historical dynamics. Drawing on mathematical models and on observations from many nations and historical empires, Turchin has concluded that the capacity of the core imperial nation for "concerted collective action" is the key factor in building large and powerful states. In other words, cooperation provides a competitive advantage and so is favored by selection among groups. However, if an empire achieves peace and prosperity in much of its territory, this means a decline in a key motivating factor for cooperation—conflict with other groups. At this point, selection within groups for selfish behavior can oppose selection among groups for internal cooperation. In addition, beyond the outer geographic edges of an empire, a different group with a strongly contrasting identity—ethnic, religious, or otherwise—might feel threatened, thus prompting a marked increase in their capacity for collective action. As a result, "the very stability and internal peace that strong empires impose contain within them the seeds of future chaos." Empires can fall rapidly.

The underlying historical models have more complexities than I have described here, but the key lesson for this chapter is that human history—just like the dynamics of ecosystems—can involve internal selective dynamics that make the evolutionary system susceptible to major and rapid changes. External factors, such as drought or other climate anomalies, can trigger or accelerate change, but the internal dynamics are what convert a potentially small perturbation into a major or critical transition. This brings us back to the global economic crisis of 2008—and similar crises of the past—which can be of a similar nature. A small increase in the stress on a system can cause widespread, rapid, and deep structural change to the entire system. The financial crisis that started in 2008 upended millions of lives and probably would have had much deeper and more long-lasting effects if not for massive bank bailouts and government intervention.

Complex Adaptive Systems

Even in simple scenarios of two competing species or technologies, we can sometimes see massive changes sparked by small pressures that push the system past a tipping point. But in most of the real-world examples, internal system dynamics are quite complex. Forests, coral reefs, lakes, financial systems, economic markets, and empires involve many interacting parts, each responding or adapting to its ever-changing circumstances. System-level characteristics, such as the dominance hierarchy of species or companies, or their total productivity or wealth, are often considered "emergent properties," in the sense that they can't be predicted just from knowledge of the individual parts. Systems of this nature have spawned an entire field of research focused on what researchers call *complex adaptive systems*. By "adaptive" they essentially mean "evolutionary," so the science of complex adaptive systems forms one important branch of the Second Science.

The dynamics of complex adaptive systems are, by their nature, difficult to predict. We can't realistically (or ethically) poke and prod ecosystems and societies in all imaginable ways to find out what kinds of pokes and prods we might need to worry about. But a generalized understanding of how complex adaptive systems work, and detailed study of specific systems, can help identify potential warning signals of imminent transitions. An anomalous increase in nonreciprocated lending patterns among banks appeared to portend the 2008 financial crisis, and a slowing down of system responses to perturbations appears to be a generic signal that an evolutionary system is approaching a tipping point. In 2010 Peter Turchin assessed economic and social indicators in the United States in relation to predictors of sociopolitical instability throughout history. Based on declining wages, increased inequality, and increasing government debts, he predicted political instability in the coming decade. He was right.

In many of the examples presented in this chapter, there was an explicit or implicit sense that major or critical transitions are things we would want to avoid. We'd rather have reefs dominated by corals, clear lakes, and thriving economies. But there are also many critical transitions that people might have an interest in triggering themselves. Some cultural norms can be severely injurious to people's health, and political polarization is a crippling state in parts of the world. Many people would very much like to see these end in critical transitions to less injurious and less polarized situations. In short, major changes are not inherently good or bad, but in either case, they can have massive consequences for humanity.

If conflict between human societies has been a major force of selection favoring cooperation within them, is it possible to achieve cooperation of *all* people on planet Earth in the face of threats—such as global climate change—that affect everyone? I certainly don't have an answer to this question, but in light of the Second Science, we shouldn't be surprised that without an adversary that is other than ourselves, the challenge is monumental. More generally, there are some undeniable ways in which decision-making and many other aspects of life have been globalized. As I look around my home and out my window, and as I take in the day's news and e-mail correspondence, it is hard not to be struck by how many products, ideas, materials, and species came from some other part of the world. Some arrived just this moment (an e-mail); others arrived several centuries ago (the dandelions in my lawn). Biological, cultural, and economic interconnectedness has reached an unprecedented level across the globe. Understanding the consequences of such globalization for evolutionary systems is one of the great contemporary scientific challenges. This is one of the central themes in the next chapter.

9

Diversity and Globalization

You can't buy a Big Mac in Antarctica, but you can buy one just about anywhere else on Earth. McDonald's restaurants are found in more than a hundred countries, on all inhabited continents, stretching from Chile, Morocco, and Finland to Pakistan, Japan, and New Zealand. The global expansion of this American restaurant chain has become emblematic of the more general process of globalization—the ever-increasing influence of distant parts of the planet on one another. Some refer to globalization as "McDonaldization." Fast-food restaurants, soft-drink brands, Hollywood movies, and "Western" values are spreading across the world, quashing local cultural diversity and pushing humanity toward cultural homogeneity.

In more than 35,000 McDonald's restaurants across the world, you will find more or less the same Big Mac sandwich, with its sesame seed bun holding two beef patties, cheese, lettuce, onion, and special sauce. Many other food and beverage companies, clothing brands, technologies, and entertainment products have similarly penetrated markets in all corners of the globe. The forces of homogenization are undeniable. At the same time, McDonald's restaurants are not actually all the same. A McDonald's menu in North America includes neither a cottage cheese and radish sandwich nor macaron cookies, both of which are on the menu in Poland. In Egypt they offer McFalafel. In Morocco you can get a chicken wrap with "Algerian sauce." In Japan you can order a Teritama burger, with pork, egg, apple, lemon, and teriyaki sauce.

In many countries, the food and ambience in a McDonald's restaurant are distinct from other offerings, and so McDonald's actually contributes to the local diversity of choices. Some local menu items wouldn't exist at all if McDonald's hadn't set up shop in so many places. I have never tried the Teritama burger, but we might think of it as a new "species" of sandwich, arising from the hybridization of the McDonald's aesthetic and Japanese cuisine. As ideas, languages, technologies, and businesses spread from one part of the world to another, there is invariably a local influence on their activities and expression, leading some people to describe globalization as "glocalization." If the example of McDonald's has you less than enthused, just think of any one of the multitude of cultural products of global fusion permeating cosmopolitan cities around the world, such as Chinese Cuban cuisine, the Jamaican patois language, or K-pop music.

The movement of people and their stuff across the globe has not only influenced the food we find in restaurants, it has changed the diversity of raw foods that go into our daily diets. In Canada in 1961, wheat and potatoes made up more than 90% of the calories derived from starchy carbohydrate sources (75% for wheat, 17% for potatoes). All other sources (maize, oats, rice, and rye) made up less than 5% each. By 2013, wheat (60%) and potatoes (11%) were still the majority, but maize (14%) and rice (13%) had increased markedly, while sweet potatoes, yams, and other roots had entered the scene as well. In other words, Canadian diets are now more diversified. And the trend is not specific to Canada or to starchy grains and vegetables: Studies covering the majority of the world's countries show consistent increases in the diversity of foods making up daily diets since the 1960s.

The movement of food plants and animals across the globe started much earlier than the twentieth century. It is hard to imagine Italian cuisine without tomatoes, but in fact, tomatoes are not native to Europe, having been introduced there from South America in the sixteenth century. Potatoes made the same journey during the same time period, and it is equally difficult to imagine the diet of eastern Europe (among other places) without potatoes. Wheat from the Middle East, rice from Asia, and maize from the Americas are now staples in human diets across the world. In many regions where particular food plants are relatively recent introductions, countless novel varieties have been developed—Roma tomatoes (United States) and Fuji apples (Japan), to name just two. Worldwide, there are thousands of varieties of both tomatoes and apples. Global movement sparks local innovation.

All this movement of domesticated plants and animals has resulted not only in more diversified local diets, but in a more homogenous distribution of raw food products across the planet. Domestic corn originated in Mexico but is now grown on all continents except Antarctica. The parallels with restaurants and other aspects of popular culture are striking. In all cases, there are strong homogenizing forces making different places on Earth more and more similar to one another in some respects, but there are also diversifying forces that increase local variety and the production of globally novel variants. In other words, we see decreased diversity *among* different places, but also increased diversity *within* those places. As we shall see later in the chapter, the same dynamic applies to wild plants and animals that have been moved—both deliberately and accidentally—across the globe. In short, the link between globalization and diversity is not as simple as one might think.

This chapter is about diversity and globalization. We will see how diversity evolves, how it varies across the globe and over time, and how it can have consequences for a variety of outcomes, such as the productivity of companies or ecosystems.

What Is Diversity?

At its core, diversity is about variety, which can have three components. First, there is the number of different types present at a given place and time, with types being anything from tool designs, toothpaste brands, or religious affiliations to genotypes, species, or habitats. This component is often called "richness" (fig. 9.1). A country with Muslims, Buddhists, and Jews is richer in religions than a country with only one or two of those, and a forest with maple and fir trees is richer in species than one with only maple (or only fir) trees.

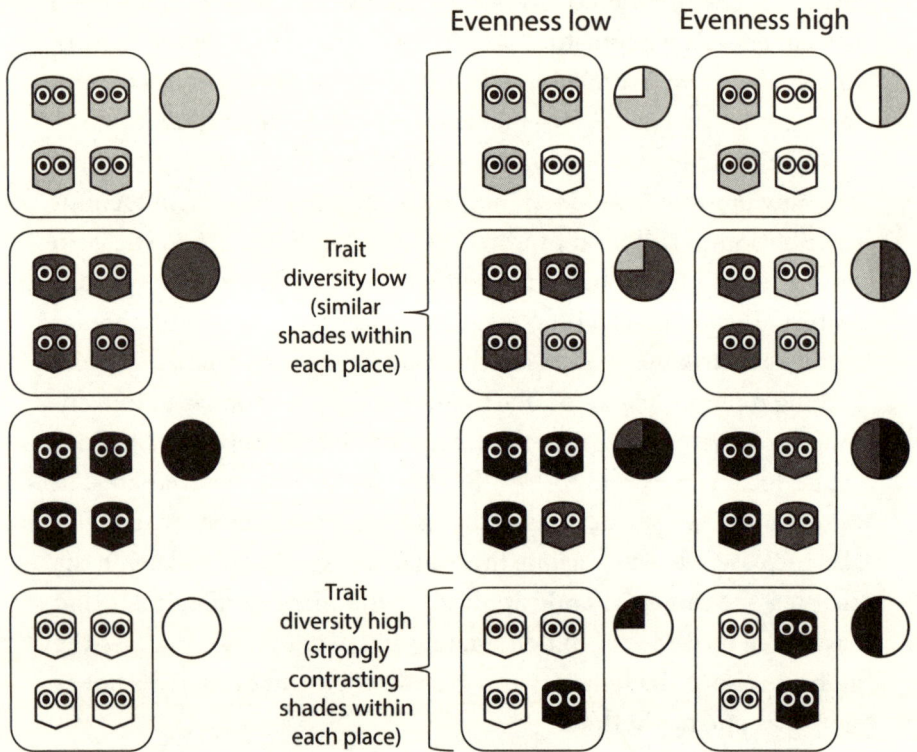

FIGURE 9.1. Diversity has multiple components: richness (the number of types), evenness (how even prevalences are across types), and trait diversity (how different local types are from one another).

The second component of diversity relates to how likely it is that a given item you encounter is going to be of a type already seen. In a forest with 99% maple and 1% fir trees, each time you look at a new tree it is almost always going to be a maple tree. The 99:1 forest is thus considered less diverse than one with a 50:50 mix, where the next tree you see is quite likely to be different from the last one you saw. This aspect of diversity—often called evenness—depends on the prevalence of each type of tree, phone, sandwich, or person in the population. In essence, when walking through the 50:50 forest, you would experience more variety moment-by-moment than when walking through the 99:1 forest.

The final component of diversity takes into account how different each type is from the others. By this criterion, a forest with fir trees and pine trees (both conifers) is less diverse than a forest with fir trees and maple trees (one conifer + one angiosperm). A menu with ten different kinds of pasta is less diverse than a menu with five kinds of pasta, two kinds of sushi, and three kinds of burrito, even if both have an item richness of 10. We can call this *trait diversity*, and it can be calculated even when there have been no predefined types. If we measure the sizes of tree leaves or the nutrient contents of fast-food meals, diversity can be calculated as the overall range of values, even if the trees or meals haven't been categorized.

Researchers in many fields have devised a mind-boggling number of indices for calculating diversity, but there is a great deal of redundancy among them. Pretty much all of them take some account of richness, sometimes incorporating evenness, trait diversity, or both.

Causes of Diversity

Evolution was defined earlier as the "trajectory of trait prevalence over time," where the trait in question can be simply the identity of a type. The state of an evolutionary system can thus be visualized as a pie chart showing, for example, how much of a plate is covered with different slime molds, the market share of different companies, the degree of support for different hypotheses (fig. 3.2), or the prevalence of different critters (fig. 9.1). Indices of diversity are numbers summarizing how much variety is in the pie chart. A pie with more slices has higher richness, a pie with more equal-sized slices has a greater evenness, and when traits contrast more strongly among slices (e.g., the shades in fig. 9.1), trait diversity is higher.

Diversity is under the influence of any and all processes involved in evolution—that is, the whole Evolutionary Soundboard. Focusing on richness (R), change over time can happen in three ways. New types can be added if they originate locally (variation generation) or if they are introduced from elsewhere (immigration). Types are eliminated when they go extinct. So, richness at some future time is equal to the

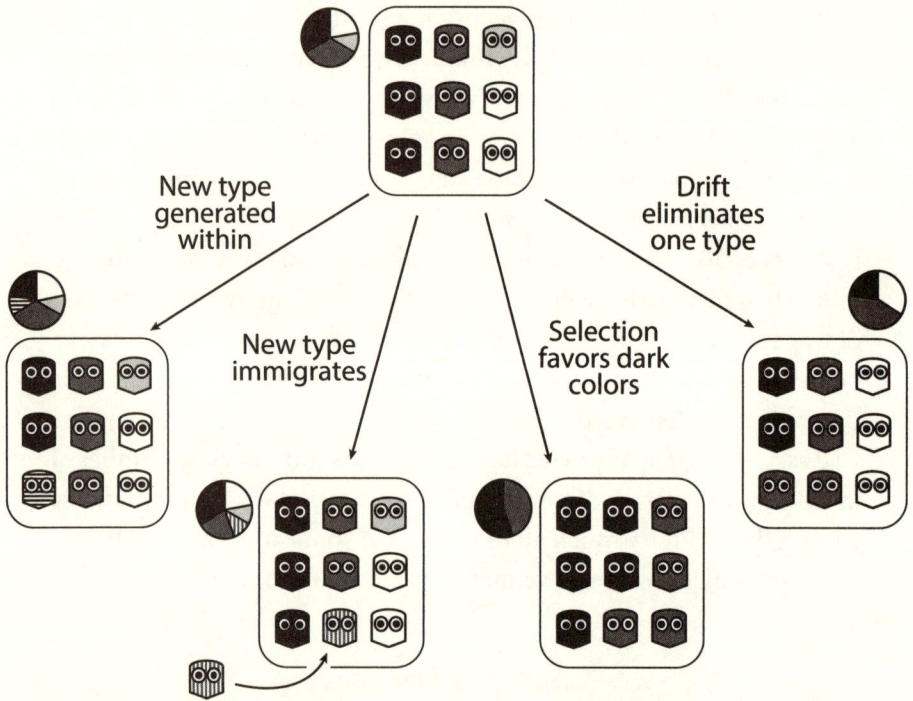

FIGURE 9.2. The influence of evolutionary processes on diversity.

present richness, plus the number of new types originating locally during the intervening time period, plus the number of types immigrating in, minus the number going extinct:

$$R_{future} = R_{present} + \text{Local origination} + \text{Immigration} - \text{Extinction}$$

Tuning up the Evolutionary Soundboard dials for the rate of variation generation and for movement (the source of immigration) should, all else equal, cause diversity to increase (fig. 9.2). When Teritama burgers were created and when Big Macs arrived in Japan, the richness of sandwich options went up.

In evolutionary terms, extinction is the end point of some combination of selection and drift. If traits caused a systematic disadvantage of certain types, then selection contributed to their extinction. If McDonald's customers dislike the flavor combination of the Teritama burger, it will likely disappear from the menu. Purely random accidents can also be involved, in which case drift had a role to play as well. When a storm randomly killed off all the beefsteak tomatoes, leaving only Romas (see chapter 6), this was a contribution of drift to extinction. So, if we tune the fitness function dial to strongly favor some types over others, or if we tune down the population-size dial (enhancing the strength of drift), diversity should decline.

The influence of other soundboard dials on diversity is indirect. Under negative frequency-dependent selection, types that become rare tend to bounce back to higher abundance instead of going extinct. So, negative frequency-dependent selection helps to *maintain* diversity. The rarity of the name Madison in the latter half of the twentieth century probably helped it rise to prominence at the start of the twenty-first century. But negative frequency-dependent selection won't by itself *create* diversity (the name Madison already existed). Positive frequency-dependent selection means that types already prevalent (e.g., particular social media platforms) have a disproportionate advantage, which can accelerate the loss of diversity and make it less likely that new types will persist. For inheritance, the effects the soundboard dials on diversity will depend on what is happening with the others. For example, if selection provides a constant advantage to some types (e.g., a new type of cell phone), increasing the number of influences on inheritance (e.g., how many people you get ideas from) can accelerate evolutionary change, and so the loss of diversity. In short, all the processes that influence evolution can necessarily also influence diversity, and in many cases we can make specific predictions about the directions of such effects.

The influence of the Evolutionary Soundboard on diversity just described assumed that the diversity was being measured in one place, potentially under the influence of immigration from some other, undefined place. Many evolutionary models are built under this assumption, for example, when an island can receive immigration from a mainland, as we saw in chapter 6. Scientists also frequently build models in which those other places are explicitly defined by their own fitness functions, population sizes, and so on, in which case diversity has at least two scales of measurement: local (within each place, usually summarized as an average) and "global" (considering all places as one unit, even if their cumulative size falls short of the actual globe). The introductory section of this chapter, and the section in chapter 6 on movement, provided some initial indications as to the potential interactions among evolutionary processes when movement links places together.

The most intuitive consequence of increasing the rate of movement between multiple places is to make those places more similar to one another (see fig. 6.6). Imagine several isolated islands, each of which is home to its own unique language, type of pottery, or species of tree. If we then move some people or seeds around the archipelago, the islands will no longer be as distinct from one another. This is effectively what happened between the first and second columns of figure 9.1: Critters were shuffled around so that each place gained some new types (increased diversity) and came to share types with other places, making the combined set of four places more homogeneous.

Movement can also prompt the variation generation dial to be tuned up, given the increased chances of recombination or hybridization between types that might not otherwise meet. This can spark the emergence of globally novel variants. Chinese Cuban cuisine wouldn't exist if not for the movement of Chinese people to Cuba, and some species of plant wouldn't exist if the movement of species among continents hadn't created new opportunities for hybridization.

Geographic Patterns of Diversity

For centuries, biologists have been fascinated by the astonishing variety of forms taken by microbes, plants, and animals in all corners of the Earth, and by the fact that some corners have a lot more diversity than others. But diversity has also been a vexing subject. On one hand, some relationships between diversity and variables like latitude or the size of an island are remarkably consistent from region to region, or from one group of organisms to the next (e.g., mammals, insects, and plants), suggesting some universally applicable laws. This optimism about finding general laws is further bolstered by findings of similar relationships for aspects of cultural diversity. On the other hand, no one can seem to agree on what the laws are. The same pattern might have multiple causal explanations, and a given causal factor (e.g., climate warming) might have highly variable consequences, increasing diversity in one place (e.g., a temperate forest) and decreasing it in another (e.g., a dryland). In this section I will first present some well-known patterns of diversity, both in biology and in culture, explaining how they can be understood using the Evolutionary Soundboard. In the next section I will delve into the specific question of how the acceleration of globalization and other human activities has caused diversity to change during recent decades and centuries.

If you travel from a place with cold winters to a place that is warm year-round, or vice versa, you will be struck by vast differences in biological diversity. For most kinds of organisms, including birds, mammals, plants, and insects, there is a steady decline in diversity as you get farther away from the equator, going either north or south. And the same relationship applies whether you're comparing biodiversity measured within small parks or across whole countries. This is known as the latitudinal biodiversity gradient, and its causes have been debated for decades.

Since most biochemical reactions speed up when it's warmer, it could be that life—including speciation—just happens faster in the tropics, such that diversity accumulates more rapidly there. This would involve tuning up Dial 1 for variation generation on the Evolutionary Soundboard. However, when using DNA to estimate the time since a given species diverged from its closest relative, recent speciation actually appears to have been *slower* in the tropics than in the temperate zone. The extinction rate must therefore be higher in the cold regions of the Earth, which is not surprising given that glacial cycles over the last two or three million years have repeatedly covered polar areas in massive ice sheets. Ice sheets presumably exert strong selection. An alternative hypothesis is based on the idea that natural enemies like pathogens and herbivores have stronger effects in the tropics. If such enemies have disproportionate effects on the most abundant of the species they attack, this will strengthen negative frequency-dependent selection, and so help to maintain diversity. This hypothesis is supported by some evidence—especially for trees.

Of the many hypotheses that have been put forward to explain why biodiversity is often highest in the tropics, none has emerged as the exclusive winner. It is clear that multiple dials on the Evolutionary Soundboard are involved.

The latitudinal biodiversity gradient is paralleled by a latitudinal gradient in human linguistic and cultural diversity. Using data from the *Atlas of World Cultures*, which includes more than 3,500 groups of people defined by characteristics such as language, beliefs, and political organization, there is a clear tendency for greater cultural diversity in the tropics. In a million square kilometers, one might find roughly fifty "cultures" close to the equator, compared to ten or twenty at temperate latitudes. This result needs to be taken with a grain of salt: There can be considerable ambiguity in drawing lines between cultural groups, with any such lines likely to be contested. However, languages can be delineated with less subjectivity, and they show a latitudinal gradient on their own. Tropical areas house more native languages, each of which is spoken in a smaller geographic area, compared to areas closer to the poles. The causes of these patterns are also the subject of active research.

Some have posited that tropical areas present more natural barriers between human populations in the form of steep mountains or deep rivers, which would reduce movement between populations, thereby permitting drift and positive frequency-dependent selection to create regional language diversity. Alternatively, more constant and productive (i.e., warm and wet) climatic conditions might increase the likelihood of long-term persistence of linguistic populations that occupy small areas, effectively by reducing the regional-scale action of drift. The data more strongly support the second hypothesis: Correlations of linguistic richness with climatic variables like temperature and rainfall are stronger than correlations with topographic variables like elevational variation or river density. These same climate variables are often among the best predictors of biodiversity, suggesting that large-scale gradients in biodiversity and cultural diversity involve some of the same dials being tuned on the Evolutionary Soundboard.

A biological pattern that is even more generally applicable than the latitudinal biodiversity gradient is the species-area relationship, by which large geographic areas house more species than small areas. Similar patterns are found for cultural diversity—for example, countries with larger populations tend to have greater religious diversity. On one hand, these patterns might sound trivial. If you grab multicolored candies from a jar, a big handful is expected to get you more different colors than a small handful, on purely statistical grounds. So, since large areas contain more individual organisms and people than small areas, they will also contain more *types* of organism (species) or people (cultural groups). On the other hand, even in this seemingly simple example, there can be at least two evolutionary processes involved. First, increasing size decreases the role of drift and so helps to maintain diversity. Second, larger areas are likely to contain a greater internal heterogeneity of environmental or social conditions. This means that there is more potential in a large area for selection to favor different types in different subsections of the area, thereby helping to maintain overall diversity. Evidence has been found for both processes at work in biological and cultural case studies, again emphasizing the conclusion that correlations between diversity and attributes of the environment can often involve tuning multiple dials on the Evolutionary Soundboard. One cannot necessarily point to a rigid law—such as a single process that always dominates—to explain repeated patterns.

Diversity Change in the Anthropocene

From Global to Local Diversity

Humans have been profoundly affecting the Earth's ecosystems for thousands of years, via the use of prescribed fire, agriculture, and hunting and gathering on land and in the sea. But the scale and magnitude of such impacts over the past few centuries seem to have crossed a line that deserves special distinction. Scientists disagree on whether to formally recognize the Anthropocene as a geological epoch, and if so, what its start date would be. Here I use the term informally to refer to the past few centuries during which human impacts on Earth have reached unprecedented levels, in the form of pollution, climate change, extinctions, and the movement of life across the globe.

The word *crisis* is frequently applied to the situation regarding both biological and cultural diversity during the Anthropocene. At the global scale, the descriptor is apt. Over millions of years, extinction has typically been a slow process. Prior to the Anthropocene, roughly one to two out of every million species went extinct per year. If we apply this extinction rate to the seventy thousand or so species of described vertebrates (mammals, birds, reptiles, amphibians, and fish), we expect roughly five to ten species to have gone extinct since 1900. In fact, *hundreds* of vertebrate species have gone extinct since 1900, indicating that during the Anthropocene, the extinction rate has increased by a factor of at least 10, and probably more like 100. The leading cause of Anthropocene extinction has been strong selection imposed by a variety of human activities: habitat conversion, pollution, overharvesting, and nonnative species introductions.

Languages have seen a high rate of Anthropocene extinction as well. A comprehensive compilation identified 7,268 languages that were used as a first language somewhere in the world in the year 1700. Almost 10% of these (701) are now extinct or "dormant" (still of some cultural significance, but without any fluent speakers). Another 1,049 languages (14% of the total) are not being learned by the current generation of children and so are considered "shifting," "moribund," or "nearly extinct." Every forty days or so, the world loses one more language. As with species, language loss is underpinned by strong selection. Historically, languages have been lost via persecution of ethnolinguistic groups, involving warfare and genocide, disease introduction, and cultural suppression. Current causes of language endangerment appear to be more related to economic development and education, which create incentives to speak widespread languages instead of highly localized ones.

These data show clear and rapid losses of biological and cultural diversity at the global scale. However, when we shift down to the scale of parks, cities, landscapes, or countries, things look quite different. In the city of Montréal, you can find roughly 200 species of tree growing— more than the total number of native tree species across all of Canada (about 140). This is not due to some anomalous quirk of Montréal. Cities are defined by their high human population densities and coverage

in asphalt and concrete, but contrary to intuition, they can also be home
to an amazing diversity of plants and animals, many of which were introduced during the Anthropocene with the globalized movement of
people across the Earth. And the diversity-enhancing effects of globalization extend beyond the city limits.

Oceanic islands have been hotspots of global species' extinctions,
including the loss of the emblematic dodo bird from the island of Mau-
ritius in the late seventeenth century. But islands are also hotspots of
nonnative species naturalizations. For every bird species that has gone
extinct on an island during the Anthropocene, there has been, on average, one new species from elsewhere that has established itself on the
same island. Plant introductions by people have outpaced extinctions
to the point that the number of plant species found on a typical oceanic
island has roughly *doubled* during the Anthropocene. Comparably sized
areas of continents (e.g., American states) show a similar constancy in
the richness of birds, and an increase in plant richness, albeit of smaller
magnitude than seen on islands. Some of the species being introduced
can drive others extinct: When predators like cats or ferrets are introduced to islands, they have often exterminated flightless birds that
evolved in the absence of such predators (e.g., kiwis in New Zealand).
In addition, in many local areas, especially those converted to agriculture, intense selection has caused dramatic losses of biodiversity. But
outside of agricultural monocultures and parking lots, local gains in species often match or even exceed the losses.

The loss of cultural diversity globally is happening at the same time
as some spectacular local increases. The people of New York City speak
more than seven hundred different languages—more than twice the
number of native languages in all of North America, north of Mexico.
New York might be an extreme example of cosmopolitanism, but a vast
diversity of languages and cultures can be found within cities and countries around the world, from São Paolo to London, Singapore to Sydney.
In my primary school in Toronto, I had classmates who were part of
first- or second-generation immigrant families from Chile, Puerto Rico,
Vietnam, Hungary, and, judging from memories and class photos, likely
other places not remembered or never revealed. In high school, one of

our favorite lunch spots was the Falafel King, which was next door to China House restaurant, and as I now take a virtual stroll in Google Street View on the same three blocks of Toronto, I see Korean, Indian, Thai, and pizza restaurants, along with a Jewish bakery and a French one as well. The source of all this diversity has been a steady tuning up of the movement dial during the Anthropocene.

Homogenization (or Not)

As we saw with the case of McDonald's and similar multinational corporations, greatly accelerated movement has led to some glaring ways in which the world is now a more homogeneous place than it used to be. If the spot I just mentioned in Toronto came up in the online game GeoGuessr (in which you guess the location of random photos in Google Street View), there would be few clues to help narrow things down further than North America. The scientific journals I contribute to are published almost exclusively in one language—English—whether they are published in North America, Europe, or Asia. Common dandelions, which are native to Europe and Asia, are now found on all continents except Antarctica, and analyses of entire floras in different parts of the world show them to have become much more similar during the Anthropocene. But as in the case of local diversity, not all Anthropocene changes match the simple expectation that the world has become a more homogeneous place.

Within landscapes or countries in many parts of the world, conversion of a portion of lands to human uses has created a heterogeneous mix of ecosystem types. For example, biologically speaking, Canada and the United Kingdom have probably become more internally *heterogeneous*, at the same time that the two countries have become more similar to one another. Both places have seen large additions of pastures, agricultural fields, and urban areas, each with a flora and fauna that is habitat-specific (i.e., different in a pasture than a crop field) but similar across countries (i.e., a similar mix of species in pastures in the two places). These habitats were added to what previously would have been mostly forest, grassland, and tundra, each with quite distinct floras and faunas across continents (except in the case of tundra, of which there is almost none in the United Kingdom anyway).

In the social sciences, there has been intense interest in the question of whether globalization—in the form of increased economic exchange and migration flows—has caused cultural homogenization or cultural differentiation (or neither). Study results are highly variable, depending on which parts of the globe are included, and what characteristics are assessed to represent culture. A study that captured the essence of a much larger literature was conducted by economist Anneli Kaasa and linguist Michael Minkov, analyzing the widely used World Values Survey. Starting in 1981, the World Values Survey project has used detailed questionnaires to assess people's values in what is now almost a hundred countries. Questions probe people's desires, morals, or opinions concerning a wide range of topics, from religion, obedience, and personal independence to homosexuality, divorce, and leisure activities. Kaasa and Minkov tracked changes between the mid-1990s and the early 2010s in a globally representative set of eighteen countries.

The answers to many questions in the World Values Survey are strongly correlated, allowing researchers to extract what is known as an axis of variation, which summarizes, in a single number, the extent to which people value personal freedom (individualism) versus the collective good (collectivism). Most countries have evolved in recent decades toward more individualistic values, characterized, for example, by greater acceptance of homosexuality and a greater emphasis on leisure and friends compared to religion and work. If all countries moved along this axis at the same rate, there would be no change in how different they are from one another—that is, no tendency for homogenization or differentiation. And this is what the study found for a subset of questions categorized as "personal values," relating to things like placing more importance on leisure and friends than religion and work. In contrast, for questions related to children—such as the importance of fostering independence and imagination over faith and obedience—there was strong homogenization. Countries where independence and imagination were already highly valued did not change a lot, while other countries shifted markedly in their direction. Finally, the opposite result—divergence—was found for questions categorized as "moral ideology," concerning the acceptability of homosexuality, abortion, or divorce. More "permissive" countries moved steadily toward greater permissiveness, while less permissive countries did not.

These contrasts reveal likely interactions between the movement and selection dials on the Evolutionary Soundboard. The shared global direction of evolution along the collectivism-individualism axis suggests a common direction of selection, which would be unlikely if not for the massive movement of people and ideas across the planet. But when new people or new ideas push against local values and norms, this can also prompt counterefforts that push in the opposite direction. For characteristics that show some initial divergence, there is the potential for polarization: the creation of groups within which there is homogenization, but among which there is enhanced differentiation.

From Local to Global Diversity

When European colonists introduced apples to North America, a new food resource became available for any number of native animals. In one species of insect now known as the apple maggot fly, some individuals switched their egg laying and feeding from native hawthorn fruits to the nonnative apples. Apples ripen earlier than hawthorn fruits, and the population of maggot flies feeding on apples has diverged genetically from the population feeding on hawthorn, leading researchers to recognize it as a new "host race"—one major step in the creation of a new species. In short, the movement of a species between continents triggered the local creation of a new incipient species, thereby increasing global richness. The more general lesson is that local evolutionary dynamics, under the influence of increased movement, can alter global diversity.

In the biological world, species introduced to new places not only present native species with a novel resource, they also themselves experience new selection pressures, prompting genetic divergence between the introduced and native populations. This is often a first step on the road to speciation. In addition, when human-assisted movement brings previously isolated plant species into contact, hybridization is a frequent result. Plants have the habit of occasionally doubling their number of chromosomes, which can essentially create a new hybrid species overnight, genetically isolated from either of the parent species. In Great Britain alone, six to seven new plant species have come into existence since 1700. This might not sound like much, but it corresponds to a rate of speciation at least fifty times that of the background rate.

Comparable things are happening in the cultural world, and as with cultural evolution in general, they can happen a whole lot faster than in the biological world. Hybrid forms of fashion, music, and cuisine arise all the time, some quite ephemeral, others more long-lasting. When peoples of different languages need to work together—frequently as slaves historically—they often communicate in a language called a pidgin, a simplified version of the local language, augmented with elements of the different languages in the mixture. Children in many such communities have transformed pidgins into what are known as creoles: full-fledged hybrid languages, with all the grammatical and syntactical complexity of the parent languages. Afrikaans evolved in southern Africa in the eighteenth century, from largely Dutch origins, with influences of German and the local Khoisan languages. Afrikaans is now spoken by millions of people. One of the youngest known languages is called Light Warlpiri, which evolved starting in the 1980s from a mixture of the indigenous Australian language Warlpiri, Kriol (itself a creole language), and English. It is spoken by a few hundred people in Lajamanu, Australia. Other examples abound.

While the global story about Anthropocene diversity is largely one of loss, some of the same evolutionary forces that cause some types to be lost can also trigger the creation of new diversity. For biological species, diversity creation probably happens too slowly to compensate for the losses. For culture, diversity can be assessed in so many ways that it is easy to imagine some ways in which global diversity has been impoverished during the Anthropocene, and others in which humanity is now as diverse as, or more diverse than, it ever has been. In sum, increased movement can simultaneously trigger increased local diversity, greater homogenization between locations, and altered global diversity as well (fig. 9.3).

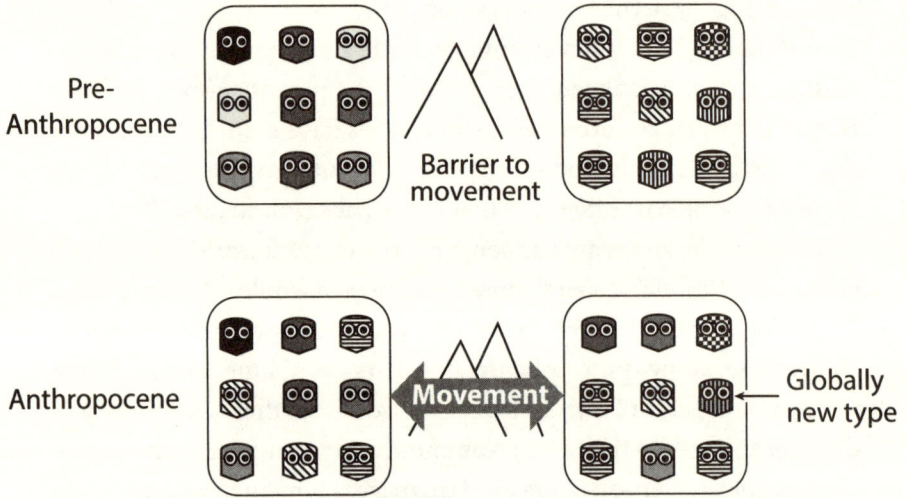

FIGURE 9.3. The influence of enhanced movement on diversity and homogenization. With increased movement, new types are introduced from one place to another (e.g., striped critters from *right to left*), the two places come to share several types (homogenize), some types can go globally extinct (e.g., the palest gray at *top left*), and globally new types can arise (e.g., via hybridization: the gray critter with vertical stripes on the *right*).

Valuing Diversity

Discussions of diversity often take on a tone different from discussions of other attributes of evolutionary systems—like the average size of cell phones or of bird beaks. Learning that phones or beaks got smaller or larger doesn't typically spark any sense of joy or dismay. In contrast, for many people, learning of changes in diversity sparks immediate value judgments. In my own field of ecology, diversity is celebrated, and facts about diversity are almost always communicated along with some value judgments: The proliferation of diversity is a good thing; its loss, a bad thing. A similar vibe permeates much of the social science literature. The next section will address some of the consequences of diversity for things like ecosystem productivity, economic growth, or the creativity of working teams—relationships that can be tested using science. To evaluate the societal implications of these studies, it is important to first appreciate potential clashes between science and values in the study of diversity.

There is no shortage of arguments in the scientific literature along the lines of "diversity is good," but in many cases, close inspection suggests that it is not diversity specifically that is being judged as good, but rather a particular *composition* of an evolutionary system. Diversity is about variety—how many types, how equitable their abundances, and how different they are from one another. Composition is about *which* types are present and their specific traits. It is possible for a system to undergo a complete change in composition without any change in diversity at all—for example, if maple and fir trees are replaced by oak and pine trees (richness = 2 in both cases, with similar trait differences). In figure 9.3, the two pre-Anthropocene places have exactly the same diversity but completely different compositions.

If it is really composition that is being valued more than diversity, how do we decide what constitutes "good" composition? In many cases, it is whatever the composition was at some time in the past—often pre-Anthropocene—and any deviations from the desired composition are judged as "bad." Basically, all change is bad. In the book *How New Languages Emerge* (2006), David Lightfoot summarized the situation nicely: "Languages come and languages go. We deplore it when they go, because the disappearance of a language is a loss for the richness of human experience. . . . Meanwhile new languages are emerging and we often deplore that, too, on the grounds that new forms represent a kind of decay and degenerate speech that violates norms that we have been taught in school." The exact same value judgments are frequently applied to other aspects of culture, and also to ecosystems, where losing native species *and* gaining nonnative species are both considered to compromise the "integrity" of the ecosystem.

If we're interested in a maximally objective understanding of how diversity has changed, a challenge arises when reports of change count losses and not gains, or when we say "diversity" but really mean some aspect of composition. For example, a quantity called the Index of Linguistic Diversity showed marked declines between 1970 and 2005 in all five biogeographic regions to which it was applied: Nearctic, Neotropical, Palearctic, Afrotropical, and Indo-Pacific. One might logically conclude that the variety of languages spoken in each region is now less than it was fifty years ago. The index, however, applies only to the languages *native* to a given region, many of which have indeed been lost, rather than accounting for both losses and gains in the languages actually spoken by people. Counting losses may well be of interest on its own, but it tells only one part of the global story of language diversity. Similarly, reports of ecological change over time often include trends in indices labeled as "biodiversity intactness." A downward trend in such an index indicates how different things have become from a (presumably desired) historical state, but this may or not be accompanied by any change in actual diversity.

In treating the subject of diversity, I have tried to avoid building in value judgments, or conflating diversity with composition. I am willing to guess that the word "McDonald's," at the start of this chapter, immediately primed some readers' minds to hear a story of evil versus good, rather than an account of ways in which multinational corporations can increase local (and even global) diversity. The point is not to argue that multinationals have been a net benefit to humanity, however one might choose to measure that. The point is to illustrate how a full understanding of how diversity varies across the Earth, and how it has changed over time under globalization, requires that we keep politics out of the measurement phase. Once we see what the data have to say, each of us can decide whether the news is good or bad.

The Consequences of Diversity: Bonus or Penalty?

Without diversity, evolution would come to a screeching halt. So, for evolution to continue, some diversity is necessary. But is more diversity necessarily better? The answer to that question depends on what is meant by "better," and even once there is some agreement on that (e.g., we might consider an efficient work team to be better than an inefficient one), the answer is sometimes "yes" and sometimes "no." Although more often yes.

Let's conduct a thought experiment, starting with a fact: The boreal forest of northern Canada has a much lower diversity of trees than the tropical forests of Brazil. Imagine that we want to promote the adaptation of northern forests to climate warming, so we choose to plant a larger diversity of trees than is already present, since diversity fuels evolution. We go get small trees of one hundred species in Brazil, and we plant them alongside the native trees in northern Canada. What happens?

You can probably already see that this is a bad idea. Our diverse experimental forest turns out to be much less productive than one where we planted only resident native species. Almost all the tropical trees die during the first winter, leaving big patches of bare soil, much of which gets carried away during summer rains, compromising the growth of the remaining native trees. There might also be conflicts between species. A few tropical trees might survive a few years (e.g., those from mountain tops) and produce a chemical that is toxic to native trees, further reducing their growth. In short, increasing diversity means adding things that are different, but if the existing types are well adapted, adding things that are *too* different is not likely to improve average fitness, either in the short term or the long term.

This hypothetical example yields two lessons: First, diversity is not always beneficial to adaptation. Second, a change in diversity necessitates a change in composition at the same time. If we add new tree species to the Canadian forest, we have changed not only how many species are there (diversity) but also *which* species are there (composition). It was impossible to change diversity without also changing composition. As we saw in the previous section, the reverse *is* possible: Composition can change—swapping maple and fir for oak and pine—without any meaningful change in diversity.

To generalize beyond trees, imagine any evolutionary system in which the fitness function dial is tuned so that optimum fitness occurs at some intermediate trait value (fig. 9.4). An example used earlier was cell phones—best not to be too small or too large. If the current average trait value is smaller or larger than the optimum (the triangle in fig. 9.4), then it is diversity that permits a response to selection: evolution of the trait toward the optimum.

But can diversity itself influence average fitness in the short term? The fitness function in figure 9.4 is not a straight line, increasing faster to the right of the triangle than it decreases to the left. This means that

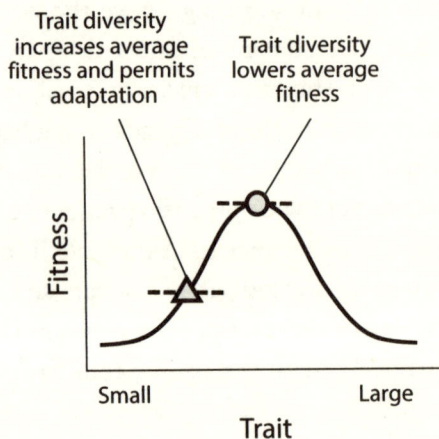

FIGURE 9.4. Scenarios in which trait diversity can enhance or diminish average fitness and adaptation. Shapes are positioned at hypothetical average trait values. Dashed lines represent the range (diversity) of trait values.

adding diversity (the dashed line, symmetric around the triangle) also increases average fitness immediately: A phone that is 1 centimeter larger might increase sales by 1,000 units, while a phone that is 1 centimeter smaller decreases sales by only 700 units, so if you add both (variation), you've increased average sales by 300 units. But if the average size is already at the optimum (the circle), then diversity actually comes at a cost to average fitness. Both smaller *and* larger phones have decreased sales. In this case, maximum average fitness occurs when all individuals have the same exact (optimum) trait value—that is, when there is no diversity at all. We can think of the average position along the trait axis (e.g., the circle is further to the right than the triangle) as trait composition, and the breadth of trait values along the same axis (the dashed lines, which are the same width in both cases) as trait diversity. Both composition and diversity make a difference.

In reality, it is unlikely that there will be one optimum trait value that always stays the same. As features are added to cell phones (e.g., enhanced video), people's preferred sizes will shift. As the climate warms, so will the locally optimal traits of trees. This means that extremely low diversity is unlikely to ever maximize average fitness over the long term, even if it does sometimes in the short term (the circle in fig. 9.4).

Evolution is constantly chasing a fluctuating environment. But as diversity increases, it matters *which* new traits are being added. Starting from the triangle in figure 9.4, adding only smaller trait values would increase diversity but cause average fitness to decline and do nothing to permit adaptation. In the case of the boreal forest, if climate warming has already decreased the fitness of resident trees, adding some species from the temperate forest just to the south (rather than from a tropical forest) might be just the thing needed to ensure long-term forest productivity. If a disease outbreak occurs, as it did for potatoes in Ireland in the nineteenth century, having a diversity of plant species, or of disease-resistant genes in a focal species (e.g., potatoes), could provide insurance against massive system-wide decline (e.g., enduring food shortage and starvation).

There is one additional reason that diversity can have short-term benefits, expressed most clearly in evolutionary systems where negative frequency-dependent selection dominates. Economic systems are a good example. Take the pork industry. If the number of farms is high enough so that the production of pigs exceeds the capacity of slaughterhouses to process them, then the success of each farmer will decline. If there are more slaughterhouses than needed given the production of pigs, then the slaughter businesses will suffer as well. For both, fitness declines with increasing prevalence, and the system as a whole is most economically productive when all the parts are present in appropriate amounts. These system parts can extend to include feed suppliers, distributors, and butchers, among others. A strikingly similar dynamic is at play in ecosystems, where a diverse mix of plant species (assuming they are all appropriate to the local climate and soil) is typically more productive in terms of biomass than a monoculture or a mix of just a couple species. In these cases, a benefit of diversity derives not from fueling evolutionary change, but from optimizing the use of limiting resources.

Because the diversity and composition of evolutionary systems— populations of humans and other species, ecosystems, businesses, and economies—have been changing rapidly during the Anthropocene, there has been intense interest in understanding the consequences of these changes. As we have seen, theory suggests that things could go either way. Increasing diversity could have positive or negative effects on system-level properties, depending on the details of underlying processes.

In ecosystems, dozens of studies have created small experimental field plots (e.g., a few square meters of prairie) or containers (e.g., miniature ponds) with different numbers and types of species. By far the most common result is for diversity to have a positive effect on the productivity of the ecosystem—that is, how much biomass is produced—and related "ecosystem functions," such as how well limiting resources like nutrients are kept in the system (fig. 9.5). Two of the mechanisms discussed so far are involved. First, diversity permits adaptation by increasing the likelihood that species with high fitness will be present. Second, it unleashes complementary interactions between species, just like those between pig producers and processors. A classic example involves pastures in which legumes, like clover or alfalfa, grow together with grasses. Legume roots have a symbiotic relationship with bacteria, which capture nitrogen from the atmosphere, adding it eventually to the soil. Grasses have a high nitrogen demand and produce lots of biomass. Together, legumes and grasses produce a lot more forage for grazing animals than either one can when growing alone. This example also is a reminder of the all-important effects of composition: It is not just diversity that matters, but which particular types are combined.

Social scientists have studied the consequences of diversity in many contexts, especially with respect to immigration and economics, and the diversification of professional teams in business and science. Bringing together people with different backgrounds, skills, and expertise can permit efficient division of labor and create synergies that spark innovation, creative solutions to management challenges, and economic growth. At the same time, mixing different languages and value systems can create major challenges to communication and coordination and erode levels of trust. These positive and negative effects of diversity—acting simultaneously—have been dubbed "the paradox of diversity."

Not surprisingly, correlations between aspects of human cultural diversity and economic performance can be either positive, negative, or

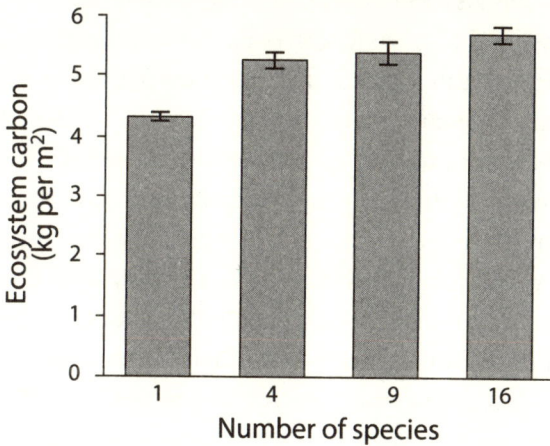

FIGURE 9.5. An experiment at the Cedar Creek Ecosystem Science Reserve in Minnesota, where the number of plant species in a grassland was manipulated and found to enhance the amount of carbon stored in the ecosystem. The image shows one of the 2×2-meter plots with one species, Kentucky bluegrass. Photo by Elizabeth Kleynhans.

neutral, depending on context. Comparing American states over the period 1960–2010, stronger economic growth was positively correlated with a greater diversity of birthplaces among immigrants, but only when looking at what the researchers described as "high-skilled" workers (those with a college education). There was no such relationship for "low-skilled" immigrants. Across African countries, ethnic diversity was found to be associated with lower economic growth, along with political instability and limited infrastructure. However, the latter result is more the exception than the rule. Reviewing dozens of studies that focus on economic innovation and productivity, within individual companies or across cities, states, or countries, researchers find positive effects of cultural diversity far more often than negative effects. Studies like these rely on correlations, and while they typically expend considerable effort in controlling for other factors, such as the total number of immigrants or population density, any conclusions of cause and effect are tentative. To some degree, however, the consequences of diversity can also be studied in controlled experiments.

The potential for diversity to create misunderstanding or communication issues, but also (and more often) bonuses in terms of enhanced problem solving, is also evident in some experimental studies. In one study, researchers tested the influence of a diversity of perspectives on the performance of management plans for striped bass fisheries in Massachusetts. First, commercial fishers, recreational fishers, and managers were asked to independently provide semiquantitative scores to causal links between striped bass abundance, environmental stressors, predators and prey, and fishing pressure. Assessments were then averaged within different groups of people—sometimes of low diversity (e.g., only commercial fishers), sometimes high diversity (e.g., a perfectly even mix of all three perspectives), and everything in between. The performance of each assessment was then rated by independent experts. Greater performance was strongly associated with higher diversity of expertise and experience. In studies of this nature, more often than not higher levels of what researchers call cognitive diversity—variation in people's process of interpretation and reasoning—yields a diversity bonus: improved performance in problem solving.

Studies on the consequences of diversity in biological and cultural settings show remarkable convergence in their conclusions. The central one is that diversity often has positive effects on productivity, efficiency, or performance (measured in a variety of ways), but not always. In thinking about the various reasons one might expect to find effects of diversity, these mixed results are not surprising. First, only certain kinds of diversity are relevant for particular tasks or outcomes. The productivity of a forest might depend on the diversity of nutrient uptake strategies among trees, but not on the diversity of flower colors. The performance of a forest management agency might depend on diversity of expertise across some knowledge areas, such as biology, forestry, hydrology, and economics, but not others, such as literature, music, or athletics. Second, increasing diversity can involve introducing types that come into conflict with others—species that produce chemicals toxic to others, or people with strongly clashing value systems. Finally, some kinds of outcomes are just not likely to be influenced by diversity, such as routine tasks like fruit picking or shelf stocking, where the group-level performance is not strongly influenced by interactions between individuals.

Some of these points might seem obvious, but they force us to be wary of any broad claim that enhanced diversity always brings net performance benefits. An increased diversity of species in an ecosystem, or of identities in a company or a society (e.g., in terms of gender, sexuality, or ethnicity), is only expected to yield performance benefits to the extent that they are associated with an increased diversity of strategies or perspectives relevant to the chosen outcome. In addition, the magnitude of diversity effects is often fairly small, relative to other factors that influence the outcomes of interest, whether in ecosystems or economies.

This chapter has revealed some striking parallels across many kinds of evolutionary system in terms of how diversity varies across space, how it has changed over time during the Anthropocene, the important Evolutionary Soundboard dials at play, and what system-level consequences one might expect as diversity changes. This makes the topic of diversity an especially potent unifying theme of the Second Science, the importance of which will only increase in the coming decades and centuries.

10

The Future
of the Second Science

At the start of this book, I described my childhood experience of reading French and English on the side of a cereal box. French and English are clearly distinct languages, but a great many people mix them together in various ways. The mixtures are sometimes given hybrid names like Frenglish or Franglais, and sometimes a new name, like Chiac, which is spoken by people of Acadian heritage in one corner of eastern Canada. Even within the bounds of one language, each speaker has their own unique linguistic expression—their idiolect—and recognizable dialects have evolved in many geographically or socially defined groups. It takes some effort (for me anyway) to recognize the English spoken in Scotland and in Jamaica as the same language. But despite all the variation and mixing, French and English (and thousands of others) are useful and important categories to recognize. Replacing school classes in French and English with one class in language would not be helpful. The point is: People like categories.

The nature of language itself—the use of nouns in particular—would seem to favor the evolution of categories. Consider the colors of the rainbow. Although a million or so colors are distinguishable by the human eye, no more than a dozen have names that are used consistently in any given language or culture. The evolution of color lexicons shows remarkable convergence across cultures in terms of which categories are recognized (e.g., black, white, and red by almost everyone), but also considerable diversity in the number of named categories (as few as two) and in the locations of boundaries between them. If you and I look at an object reflecting radiation at a wavelength of 590 nanometers, I might call it yellow while you call it orange, but without a handful of words like yellow and orange it would be difficult to communicate. So, it's not just that people *like* categories—they probably *need* categories.

This book has been an extended argument for the importance of recognizing a major category of science—the Second Science—based on a distinct set of processes that underlie how everything evolves, from cells to cell phones, from proteins to politics. The Second Science obviously fits within an even broader category: science. As an approach to knowledge creation, science is distinguished by what the philosopher of science Michael Strevens calls its "iron rule of explanation." According to the iron rule, competition between ideas is ultimately resolved by empirical evidence and nothing else, even if other factors, such as religion or particularly persuasive individuals, can have a temporary influence. Science is a descendent of "natural philosophy," having taken its current name only quite recently, in the nineteenth century.

Within science, the dominant "deep" categorical split at present, at least in North America, is between the natural sciences and the social sciences. Natural science is typically understood to include physics, chemistry, biology, and all related disciplines (geology, astronomy, physiology, etc.) aimed at explaining how the natural world works. Social science aims to understand human behavior and the outcomes of social interactions and includes the subdisciplines of economics, political science, sociology, and history. These categories are essentially the same as those that appear in one of the early efforts (from the 1930s) to map the relationships among different branches of science (fig. 10.1).[1] A similar structure emerges from modern mappings based on how scientists draw on one another's work.[2]

The natural-social divide is reflected—and probably magnified—by a great many institutions that fall on one side or the other, including research-funding agencies and university faculties. As a member of a Faculty of Science (the "natural" part is implied), I cross paths in my daily life with many natural scientists, from various departments. We

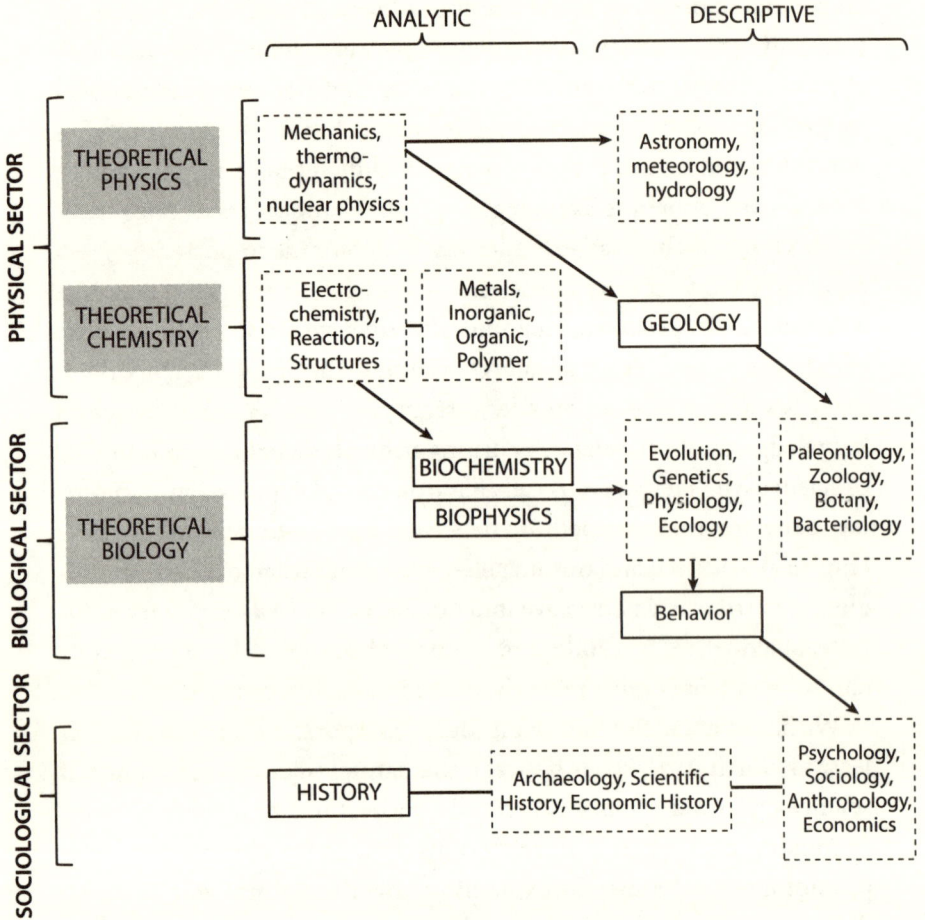

FIGURE 10.1. Simplified version of John Desmond Bernal's 1939 map of fundamental science. Arrows indicate causal influences; arrowless lines indicate connected or allied disciplines.

work in close proximity. If I want to connect with social scientists, it requires effort: They work in different faculties, in different buildings, on the other side of campus. In short, administrative and institutional boundaries channel the flow of interactions and ideas.

As a first deep categorization in the scientific enterprise, the natural-social distinction is not the only option. The topic of people and their social interactions is fascinating and important to understand. As humans, it's easy to justify investing more collective resources in this topic than, let's say, social interactions among insects. But surely one aspect of one species' existence on one planet does not make up a logically coherent category when its sister category includes all the other matter and life in the universe. A more fundamental and logically coherent distinction is between phenomena that can be explained, ultimately, using only the laws of physics (the First Science), and those that require the addition of evolutionary processes (the Second Science). This view makes one wonder if the natural-social divide is a bit like the QWERTY keyboard: It may have evolved for good enough reasons, but it hangs on, despite a better option, either because people are unaware of the better option, or because change requires more effort than we're willing to expend. We are locked into a suboptimal situation.

I am not the first person to highlight physical and evolutionary processes (or something similar) as *the* fundamental pair in science. Graham Bell coined the term *Second Science*, saying that "no knowledge of physical principles, no matter how profound or detailed, can lead to any understanding of evolution, and vice versa" (quoted also in the preface). Pioneering evolutionary biologist Ernst Mayr argued that biology is unique in large part because of "dual causation," involving not only "natural laws" (physics) but also "genetic programs." David Sloan Wilson recognized two "reasoning modes" that span the natural and social sciences: physical and functional. Things that have evolved—whether a kitchen tool or a human heart—are impossible to understand without considering the *function* served by whatever traits were retained by selection. In *A World Beyond Physics* (2019), Stuart Kauffman argued that "physics will not tell us whence we come, how arrived, why the human heart exists, nor why I can buy nectarines in Eastsound." Evolution tells us those things. Directly or indirectly, these views all argue that when systems gain the capacity for evolution, something has happened that is more profound than the emergence of a bipedal, big-brained animal that does lots of socializing. That is, the physical-evolutionary distinction is more meaningful than the natural-social one.

Given the decades-long history of recognizing the Second Science (or something like it), why does it remain on the margins of scientific teaching and discourse? By the time my children became teenagers, they had a sense of the differences between physics, chemistry, biology, and social studies, in the same way that their great grandparents might have (fig. 10.1). The undergraduate students in my classes grasp the same distinctions, and they know that the social sciences are something they mostly left behind when they joined the Faculty of Science. Social science happens on the other side of campus. But if science has led us to the insight that the universe is governed by two major classes of process—physical and evolutionary—that surely seems like something that *everyone* should learn the moment they dip their toes into the world of science. But to get there, some important roadblocks to the development of the Second Science need to be overcome.

Overcoming the Darwinian Distraction

Concerted efforts to build a unified evolutionary science have been hobbled, in one way or another, by the Darwinian Distraction, which was introduced in chapter 1. Many proponents of generalized evolutionary theory have wedded their project too closely to a "Darwinian" subset of evolutionary processes, in which variation is produced blindly, inherited only via genes or something like genes, and sorted by natural selection.

Charles Darwin and anything bearing his name seem to have become unimpeachable in the eyes of biologists and allied scientists, such that admitting non-Darwinian processes into the fold feels something like failure or disrespect. And even when a narrow focus on Darwin is recognized as counterproductive, the related tendency to assign namesakes to complex sets of ideas persists. For example, cultural evolution has been described as involving "Darwinian nested within Super-Lamarckian inheritance." A biologist can almost get the meaning of this, but the commonalities among evolutionary systems can be seen much more clearly when focusing on the fundamental processes represented in the Evolutionary Soundboard, rather than on people's names. Given the diverse and deep-rooted strands of evolutionary thinking over the centuries, I see the tying of generalized evolutionary thinking to one or a few specific names as having far greater costs than benefits. Names are important for understanding history, but not necessarily for understanding how things work.

Setting aside the Darwinian label for all things evolutionary might also help draw a cleaner distinction between social Darwinism and evolutionary science. Although you would be hard pressed to find respected scientists at present who claim support for racist and sexist ideologies from evolutionary science, the associations linger. Simply calling evolutionary science something other than "Darwinian" doesn't make the problem go away, but it can be one step in a process of separating the question of what evolutionary science has to contribute to understanding the world from the question of how human beings ought to treat one another. Indeed, social Darwinism is just one example of a more general tendency for people to see and promote whatever implications from science that align with their politics and values. The problem is that with just about any complex societal issue, one can find bits of science that appear to align with one's value position, whatever it is.

Contemporary arguments in favor of generalized evolutionary thinking can veer explicitly into politics and values, even if they avoid or reject social Darwinism. Two notable books, *The Evolution of Everything* (2015), by Matt Ridley, and *This View of Life* (2019), by David Sloan Wilson, argue for the broad application of evolutionary theory to biology and culture, but with strongly contrasting political arguments. *The Evolution of Everything* describes the best products of culture—everything from economic and legal systems to educational and government institutions—as having arisen when evolution was allowed to run free. When individuals are free to pursue their self-interest, things evolve in a way that most often translates into greater common good: laissez-faire economics writ large. Similarly, *This View of Life* "unabashedly goes beyond what *is* to provide a blueprint for what *ought to become*," but with a very different take on what is good and how things ought to be. This time, evolutionary science is seen as supporting the need for greater coordination and cooperation among people—basically *less* individual freedom to pursue self-interest—to achieve a better global society. Both books present compelling examples to support their points, but in the end, the core arguments appear to be mostly about how things ought to be, a question on which science—evolutionary or otherwise—is silent.

In this book, I have deliberately steered clear of any implication that evolutionary science aligns with my political leanings, or with anyone else's. Given the human impulse to do otherwise, I have probably strayed from that goal at times, but keeping science and values as distinct as possible is an ideal that I think is worth striving for. Whatever your political leanings, electrons will still be attracted to protons, and life and culture will continue to evolve based on inputs from the Evolutionary Soundboard.

The Faculty of Evolutionary Sciences

If the distinction between physical and evolutionary science is cleaner and more meaningful than the distinction between natural and social science, perhaps this should be reflected in the structure of scientific institutions—from school curricula and textbooks to university faculties and scientific funding agencies. For as long as there has been a scientific evolutionary account of life on Earth, its teaching to schoolchildren has been hotly debated. The Second Science expands this debate. It has long been argued that sidelining evolution in school curricula denies students knowledge of the only explanation for life on Earth that is supported by evidence. The Second Science tells us that it also deprives them of understanding the fundamental set of processes that underlie not only life, but also the cultures and economies (and education systems) in which they live and work.

In colleges and universities, building a Faculty of Evolutionary Science alongside a Faculty of Physical Science would involve not only reshuffling departments, but actually splitting some apart. Despite doing more or less the same thing for the past twenty-five years of my professional life—teaching and doing research in the field of ecology—I've done it in university departments with four different names: Botany, Zoology, Biology, and Ecology and Evolutionary Biology. In the mid-twentieth century, many universities in North America would have had Botany and Zoology departments, and for good reason. At first glance, perhaps the most striking distinction in the living world is be-tween critters with brains and eyes that can run or fly or swim (animals) and everything else (plants, fungi, and other microbes). With the ad-vent of molecular biology, that distinction started to look less signifi-cant. All life was found to be united by the same basic functions of DNA, RNA, and proteins, and some "botanical" organisms like fungi turned out to be more closely related to animals than to green plants. Slowly but surely, most Botany and Zoology departments were dissolved. Some universities lumped all of it into Biology departments, while others kept things divided, but along different lines.

The typical current split in biology divides molecular and cell biologists from ecologists and evolutionary biologists, and it applies informally within Biology departments as well. As with most dividing lines, this one is porous, in that cell biologists might want to know how the bio-chemical makeup of a cell membrane evolved, and evolutionary biolo-gists routinely study molecules like DNA. But at its core, the distinction is meaningful. On one side, the interest is largely in *how things work* at a molecular level. The inside of a cell is a highly complex soup of mole-cules, interacting in myriad ways, and those interactions are governed almost entirely by the laws of physics.[3] For the most part, molecular biologists draw on principles of the First Science. On the other side, the interest is in *how things come to be*. Why cells contain the things they do (as opposed to other things), and how the world ended up with massive assemblages of cells that make up a human body, are questions that require evolutionary principles to answer. Cells and organisms and eco-systems *obey* the laws of physics, but they are not wholly *explained* by them. Explaining how living things come to be requires the Second Science. Basically, the distinction between the First Science and the Second Science cuts straight through biology.

Viewed from the social sciences, the Second Science project might appear to some as an attempt to have the entirety of the social sciences swallowed up by biology—the evolutionary part of biology anyway. I hope this book has demonstrated that this is *not* the case. It is a quirk of history—cultural evolutionary history—that the basic principles of evolution have come to be associated most closely with biology. It could

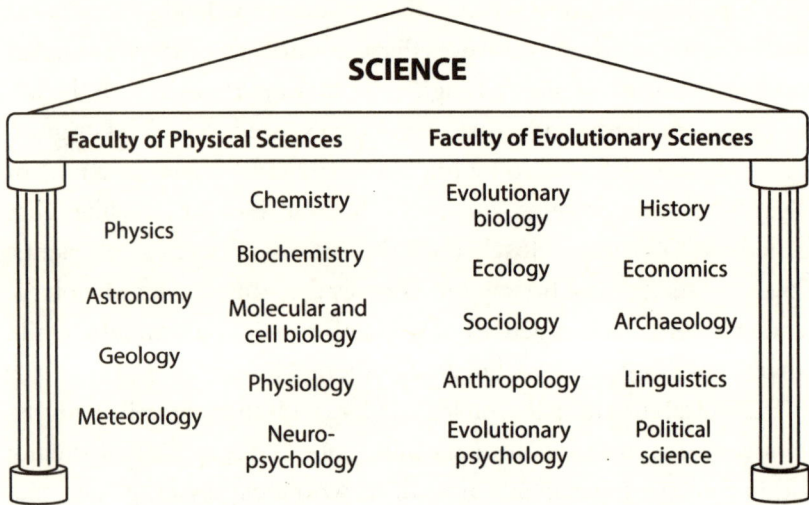

FIGURE 10.2. The future of the scientific academy.

easily have been linguistics or history. If anything, the creation of a Faculty of Evolutionary Sciences would result in evolutionary biology and ecology getting swallowed up by the social sciences. Admittedly, it would probably be wise to come up with a word different from *evolution* to more formally debiologize the concept, but I have nothing better to offer at the moment. We can wait and see what evolves.

Restructuring scientific institutions, as with any kind of social change, involves benefits and costs. Naming and recognizing new categories automatically creates boundaries between them, thereby facilitating certain connections (the benefit) while weakening others (the cost). In a Faculty of Evolutionary Sciences, I would be more likely to cross paths with linguists and economists, and less likely to see my colleagues in physics and chemistry (fig. 10.2). It seems to me like an experiment worth trying.

Choosing to See Commonalities or Differences

Pioneering evolutionary biologist Theodosius Dobzhansky famously said that "nothing in biology makes sense except in light of evolution." Biologists generally accept evolution as the unifying theoretical core of their discipline, which might lead you to expect that their day-to-day research makes constant reference to evolutionary principles. That is not quite the case.

If I walk down the hallway from my office, I will find people examining how plants defend themselves against pathogens, how agricultural practices influence the reproductive success of birds, how climate change might alter the distributions of tree species, and how much of a bacterial genome can be deleted without killing the cell. Each of these topics, in turn, requires paying attention to myriad system-specific details: defense genes in tomato plants, the timing of bird migration, the dispersal of seeds, or the resources required for *E. coli* to survive, to name just a few. Although some of this work is explicitly focused on evolution, much of it is focused on the mechanisms *underlying* evolutionary processes, without ever making explicit reference to evolution. In pondering all these systems, we could choose to see only their differences, obvious as they are. But we can also choose to see their commonalities: All the details are both inputs that move dials on the Evolutionary Soundboard and outputs of the evolutionary process.

I am not a social scientist, but it is easy enough to see the exact same tension in the social sciences between focusing on system-specifics details and identifying unifying commonalities. On the surface, there does not appear to be a lot in common between people studying the economic consequences of a minimum wage, the use of English words in modern French, or the decline of the Roman Empire. And indeed, as far as I can tell, these kinds of studies mostly proceed in isolation from one another. Even within disciplines like history or archaeology or anthropology, the postmodernist view resists attempts at finding general explanations for different systems, favoring a focus on unique, system-specific narratives. But as I hope to have demonstrated in this book, all these topics—the biological and the social—have a great deal in common, via the Evolutionary Soundboard. Systems change and take their shapes because new variants are produced, inherited, and moved around, while some variants achieve greater success than others.

The key point of this section is that embracing commonalities means neither that we should abandon the details, nor that people focusing on the details are doing it wrong. Great contributions to knowledge are made both by reporting meticulous observations—of the spiritual customs of the Inuit, of the purchasing habits of eighteen- to twenty-four-year-olds, or of the web-spinning virtuosity of spiders—*and* by examining the evolutionary significance of these observations.

Some see the system-specific, detail-oriented approaches as being in competition with those emphasizing commonalities across systems.[4] Focusing on commonalities often means that we attempt to identify a manageable number of measurable traits from a myriad of details. We might set aside the beauty and specific raw materials used to make an axe or a spider web and focus instead on their length, width, sharpness (for an axe), and internal pattern of silk strands (for a spider web). The reality is that the two approaches go hand in hand. Understanding the evolutionary pathways leading to a specific cultural practice, technology, or morphological structure provides much of the empirical evidence that defines science. At the same time, our appreciation of the details themselves can be deepened by understanding the universal processes that brought them into existence in the first place. The same is true of the physical sciences. An engineer can make painstaking measurements of the structural properties of five kinds of concrete, which are of great use in making a choice, while also aiming to understand the basis of the differences in terms of universal physicochemical laws.

Did the Second Science Come First?

The Second Science is defined by the idea that it goes beyond the First Science—physics—which has achieved startling success at revealing the nature of matter and energy. Some of the ideas in physics sound outlandish at first, like the one where the teeniest bits of matter are both particles and waves at the same time. When these wild-sounding ideas join together, not only can we explain our observations of the physical world, but we can make things like computers and lasers. Wow.

But physics can't explain everything about the physical world. The mass and charge of fundamental particles like electrons or quarks have precise values that we just need to take as a given. In our universe, what physicists call dark energy appears to be present in just the right amount to permit the existence of things like planets and galaxies. If you add a bit more dark energy, clumps of matter (e.g., planets) would get blown apart; a bit less, and everything would collapse back to whatever was there just before the Big Bang. In essence, slightly different values of these physical constants and you don't get planet Earth with its trillions of living inhabitants. But where do these values come from? No one knows, but according to one controversial theory, they might have evolved.

Our universe began with a Big Bang, with matter expanding out from one dimensionless point, and according to one theory, it will end with a Big Crunch, when everything disappears into something like a black hole. The Bang-Crunch symmetry suggests that the black holes that astronomers infer from observations of our universe might represent the starting points of new universes. The Big Bang and Big Crunch of our universe might be just one among a great many. And maybe when new universes get created, there are slight tweaks to the physical constants. If indeed there is a multiverse—a population of universes—the ones with parameters tuned just-so to permit the creation of lots of black holes, and therefore lots of new universes, will dominate.

The idea of the multiverse should sound familiar to a Second Scientist. Universes vary in their physical characteristics, which can be inherited by other universes, some of which give rise to more future universes than others. That is evolution, which makes it not so unexpected for us to have found ourselves in a universe where the fundamental physics seems so implausibly well tuned to build stars and planets. It's not that the one and only universe in existence happens to have the just-so conditions needed to support life on some of its planets. Instead, of the countless universes out there, ours is one of the few where life was even possible. If the conditions were much different, we wouldn't be here to know it one way or the other, or to create comic books and movies about multiple universes.

The multiverse idea is controversial in the scientific community, but it is out there, and, as far as I know, it hasn't been disproven yet. In evolutionary terms, it remains a viable variant in the evolution of the theory of the universe(s). And if it turns out to be true, everything we know about physics is in fact the result of an evolutionary process. In the end, maybe the Second Science is really the First.

The Future of the Second Science

Establishing the Second Science as one of two main branches of science can deepen our understanding of all evolutionary systems, the importance of which for humanity cannot be overstated. This book has revealed a common set of evolutionary processes at play in the breeding of plants and animals for the human food supply, in developing artificial intelligence, and in advertising and marketing. Evolutionary principles were shown to be central to our understanding of how and why biological and social systems have tipping points, how biological and cultural diversity are distributed across the globe, the consequences of diversity, and how all these systems respond to the pressures of the Anthropocene. Taking things a step further, each of the individual systems discussed in these chapters—particular cultures, technologies, economies, or ecosystems—interacts in a complex web of interactions that researchers are just beginning to grapple with. A unified Second Science can help us understand how these interlinked Evolutionary Soundboards combine and feed back on one another, helping anticipate the possible futures that await humanity.

NOTES

Chapter 3

1. Imagine that for a given gene, you have a rare pair of alleles that gave you bushy eyebrows, just like the two recessive W alleles that gave Mendel's pea plants their wrinkly seeds (see chapter 2). Among all your known ancestors, even if there is just one great grandparent who had the same bushy-eyebrow genotype (WW), it is entirely possible that you got your two W's from two *other* great grandparents. There are eight great grandparents to choose from, and the two you got your W's from might have each had just one W, and so they had regular-sized eyebrows.

2. I am unaware of any previous publications that present something equivalent to the Evolutionary Soundboard, but in a similar spirit, Godfrey-Smith, *Darwinian Populations and Natural Selection* (2009), compares (potential) evolutionary systems based on continuous variables such as the fidelity of heredity, the dependence of fitness on intrinsic properties, and the smoothness of the fitness landscape.

3. Godfrey-Smith (2009) argues that many putative cases of cultural evolution are "trivial" or "marginal" examples of *Darwinian* evolution. His "paradigm" version of Darwinian evolution is biological. My emphasis is on how all such examples fit the definition of evolution, regardless of how specifically Darwinian they are, and without any need to rank them with respect to a benchmark paradigm.

4. Multilevel selection and evolution have been debated for decades. Okasha, *Evolution and the Levels of Selection* (2006), makes clear that the debate is not about whether evolution at multiple levels is possible (it is) or can happen in real systems (it does), but rather how important it is—that is, how often understanding evolution at one level requires simultaneous knowledge of evolution at other levels. My concern in *Everything Evolves* is with theoretical possibilities that can't be ignored a priori, rather than how important they are in some average sense.

Chapter 4

1. In this section there are two issues of potential concern to an academic reader, but perhaps not to a general reader: blending versus particulate inheritance and the concept of heritability. As explained in chapter 2, the theory of evolution by natural selection faced a serious challenge in the nineteenth century: If offspring have traits that are intermediate between their two parents, the variation needed to fuel natural selection would quickly dissipate. For genetic inheritance, this was resolved by the discovery that offspring inherit discrete genes from each parent. For

cultural inheritance, much less is known about the physical basis of inheritance, but it is clear enough that multiple cultural influences rarely blend completely. As one example, fusion cooking and music always retain recognizable features of their "parents," rather than going halfway on each flavor, technique, instrument, or sound. As Matt Ridley, in *The Evolution of Everything* (2015), put it, the inheritance part of evolutionary change does not require strict particulate inheritance, but only "some lumpiness in the things transmitted" and "some fidelity of transmission." All evolutionary systems appear to meet these criteria.

The concept of heritability recognizes that not all trait variation is heritable. For example, despite a very high fidelity of cultural transmission for the assembly of a given cutting tool, the actual traits of the tool (e.g., strength, sharpness) can depend on the local availability of materials, such as different types of wood or rock (Lycett et al., "Factors Affecting Acheulean Handaxe Variation," 2016). If the sign in front of the first McDonald's restaurant had its "s" knocked off, the next restaurant would not be called McDonald. If I lose a finger in an accident, there will be no impact on the number of fingers my children have. To account for nonheritable trait variation, evolutionary biologists define a quantity called *heritability*: the proportion of variation in trait values among organisms that is caused by genetic differences (Wray and Visscher, "Estimating Trait Heritability," 2008). However, some of the remaining "nonheritable" variation may in fact be heritable via one of the other pathways: epigenetic, behavioral, or symbolic. The Second Science leads us to focus on the heritable component of whatever traits are of interest, via any mode of inheritance.

2. To mathematically describe a bell-shaped curve like the one in figure 4.2, we need to specify at least two parameters: the mean (where the peak falls on the x-axis) and the variance (how wide the curve is). Curves with more complex shapes require additional parameters. As a consequence, instead of just one dial (Dial 2), we could imagine adding multiple dials to the soundboard to characterize the shape of the trait distribution. The one dial, however, captures the essence of a single conceptual input into an evolutionary model.

3. Some models of the simultaneous coevolution of genes and culture also include a category labeled "oblique" inheritance, coming from an individual in someone's parents' generation that isn't actually their biological parent. In addition, as mentioned in chapter 3, the category "horizontal" might also include inheritance flowing from younger to older entities. As for some other dials on the Evolutionary Soundboard, Dial 3 stands in for some additional complexities.

Chapter 5

1. In addition to gametes, another exception to the rule of there being two copies of each chromosome in each human cell is red blood cells, which have no chromosomes at all.

2. The influence of evolutionary systems on the environments they occur in is often treated under the heading of "niche construction" (Odling-Smee et al. 2003).

Chapter 6

1. For the coin-flipping example, with a 50% chance of either heads or tails, the probability of N flips being all heads (or tails) is 0.5^N. For 200 flips, this gives 6.22×10^{-61}, and twice that (once for heads, once for tails) is 1.24×10^{-60}.

2. If we alter our perspective to consider not one island but many, the link between drift and diversity is somewhat different. In one garden, a storm might leave behind only beefsteak tomatoes, while in another garden it might be only Romas that survive (see fig. 6.1). If we started with five varieties instead of two, we still expect drift ultimately to lead each individual garden to have just a single variety (it might take several years to get there), but there is no way to predict in advance which one. So, if there are 100 independent little gardens starting with 10 plants each, it is almost certain that all the initial five varieties will come to dominate at least one little garden. In fact, if we take the same 1,000 plants that initiated the 100 gardens and put them in one big garden, drift will eliminate diversity *faster* there than it would in the 100-garden collective. The big garden is like putting all our eggs in one basket, while the little one is like spreading them among many. Each little garden is protected from competition with the others, so if beefsteaks dominate one, Romas another, and cherry tomatoes a third, each will stay that way indefinitely, as long as the population doesn't go extinct or do any mixing with the others. McShea and Brandon, in *Biology's First Law* (2010), argue for a "zero-force evolutionary law" by which diversity accumulates by random change in "independently evolving entities."

3. Translations were done at DeepL (https://www.deepl.com) in February 2023; the English-Chinese translations went into an infinite loop, flipping between two versions—here I chose one of them. The emoji translation was done at Emoji Translate (https://emojitranslate.com).

4. The International Union of Geological Sciences rejected a proposal to formally recognize the Anthropocene as a new geological epoch (Witze, "It's Final," 2024), but the term is already in widespread use in scientific journals and books and is widely understood as referring to the current period of time during which humans have had, and continue to have, massive impacts on physical and biological characteristics of the Earth.

Chapter 7

1. What I refer to as "neurons" in a neural network are usually called "nodes" by computer scientists.

2. My rough estimate of how much space would be required to write down the model underlying ChatGPT was generated as follows: I can type 72 zeros on one line of a page, which is 15.23 cm wide (not including margins). $(15.23 / 72) \times 175$ billion $= 37{,}017{,}361{,}111$ cm $= 370{,}173$ km, and that's just the characters in a line without spaces, so the equation would be longer. The moon is 384,400 km from the Earth.

3. In computer science, parameterizing models is technically done by finding a minimum on a "cost" surface, rather than by finding a maximum on a fitness landscape. The computer science approach involves calculating "gradients" and using "gradient descent" to go downhill. As the cost surface is essentially an inverted fitness landscape, I use the latter to reveal the equivalence with other evolutionary processes.

Chapter 8

1. The account of the 2008 financial crisis presented in this chapter is not meant to be definitive or authoritative, but just to illustrate one example of how evolutionary systems can change states very rapidly, with some hints as to why. For broader perspectives, see Haldane and May,

"Systemic Risk in Banking Ecosystems" (2011); Steger, *Globalization* (2020); and Buchanan, "Science Strengthened Banks" (2023).

2. Although evolved complexity makes reversal of major transitions unlikely, there are exceptions. For example, some extant unicellular fungi and algae evolved from multicellular ancestors, and some people view cancer cells (e.g., in humans) as having evolved unicellularity (Grosberg and Strathmann, "The Evolution of Multicellularity," 2007); multicellular sponges can reassemble after passing through a sieve that breaks them into their component cells (Lavrov and Kosevich, "Sponge Cell Reaggregation," 2014).

Chapter 10

1. Figure 10.1 shows only the "fundamental research" portion of the original diagram from Bernal (1939), not including the "technical research" portion (e.g., engineering, agricultural, and medical research). The original diagram also had more separate boxes (those with dotted lines in fig. 10.1 were shown as several adjacent but separate boxes), and many more connections between individual boxes.

2. "Maps" of science based on how scientists draw on one another's work, showing the strength of connections between different branches, can be produced on the Emerging Technology Observatory website, https://sciencemap.eto.tech.

3. Almost all of what happens inside a cell is governed by purely physical processes, but some evolutionary dynamics inside cells or bodies is possible. Some regions of the genome can replicate and insert themselves into other parts of the genome, as described in chapter 3. The human immune system also involves differential rates of production of different antibodies based on exposure (Cziko, *Without Miracles*, 1997).

4. System-specific descriptions are sometimes referred to as "thick descriptions" by anthropologists, as "interpretive" studies by archaeologists, or as "natural history" by biologists. In all cases, there are different approaches to these fields that focus more on commonalities, and there has been tension between the practitioners of each.

SOURCES

Preface

The observation of coevolution of the virus causing Covid-19 and societal responses is also made by Blute (2020). The books mentioned are Page (2010), Ridley (2015), Vellend (2016), Bell (2007), Blute (2010), Cziko (1997), Hodgson and Knudsen (2010), Jablonka and Lamb (2014), Mesoudi (2011), Tang (2020), and Wilson (2019). Merchant (2017) describes the history of the iPhone.

Chapter 1

Introductory Text

The section on evolutionary ideas in the history of linguistics is drawn largely from Van Wyhe (2005), augmented by additional sources (Zipf 1932, Piantadosi 2014, Campbell 2020). The quote is from Jones (1799, 26).

An Imaginary Scholar

The idea of an imaginary scholar is similar to one presented in Godfrey-Smith (2014, 48–49): "There is an alternative history, another possible world, in which Darwin or someone was able to draw on an understanding of variation and selection in a range of more obvious areas and apply it to the less obvious case of biological change." The brief history of evolutionary ideas inside and outside of biology comes from Smith (1776), Lamarck (1809), Darwin (1859), Alter (2003), Beinhocker (2006), and Gribbin and Gribbin (2022). The Metcalfe quote is from Metcalfe (1998, 36).

The Darwinian Distraction

The theoretical and mathematical synthesis in evolutionary biology was described by Huxley (1942) as the "Modern Synthesis" and is reflected in all textbooks in evolutionary biology (e.g., Futuyma and Kirkpatrick 2017). Social Darwinism is described and analyzed in detail by Paul (2009), Dennett (1996), Wilson (2019), and many others. Many books or articles extend evolutionary theory beyond biology and have titles or themes that focus on Darwin and Darwinism: "Universal Darwinism" (Dawkins 1983), "Generalized Darwinism" and "Darwin's

Conjecture" (Hodgson and Knudsen 2010), "Darwin's Dangerous Idea" (Dennett 1996), the "Second Darwinian Revolution" (Cziko 1997), "How Darwinian Theory Can Explain . . ." (Mesoudi 2011).

Two Kinds of Science

Bell (2007) introduced the term "the Second Science," and Kauffman (2019) expresses a similar sentiment. Halley's calculations are described in Cook (1998). The SARS-Cov-2 virus is described in Wu et al. (2022).

The Second Science Is Everywhere

Agricultural statistics come from the United States Department of Agriculture (USDA, National Agricultural Statistics Service 2023). Orgel's quote can be found in Dunitz and Joyce (2013).

Chapter 2

Introductory Text

Portions on the philosophy of science are drawn from Popper (1959) and Strevens (2020). The idea of science as an evolutionary process is fully developed in Hull (1988). Quoted words or phrases attributed to Whewell are from Whewell (1840). The quote arguing that Darwin's "contribution was one link in a chain that extends back into antiquity, and is still being forged today," is from Gribbin and Gribbin (2022).

The Deep Roots of Evolutionary Ideas in Biology

The main sources for the history of evolutionary biology, including the influence of religion, are Mayr (1982), Stott (2013), New Scientist (2017a), and Gribbin and Gribbin (2022). Dates of individual lives were verified using the *Stanford Encyclopedia of Philosophy* (https://plato .stanford.edu) or *Encyclopedia Britannica* (https://www.britannica.com). The interpretation of Lamarck's ideas (Lamarck 1809) as a version of the variation-differential success-inheritance recipe is from Tang (2020). Additional references include Matthew (1831, quote from 364–65), Darwin (1859, quote from 490), Mayer (1987), and Van Wyhe (2007).

The Deep Roots of Evolutionary Ideas in the Social Sciences

Some sources here are specific to particular disciplines, including linguistics (Alter 2003, Van Wyhe 2005) and economics (Smith 1776, Schotter 1981). The idea of "design without a designer" is attributed to Paley (1802), who argued that design required a designer (i.e., God). General accounts of evolutionary processes underlying economics and technology are provided in Basalla (1988), Beinhocker (2006), Arthur (2011), Ridley (2015), and Heinrich (2016). The Gould quote is from Gould (1980, 68). The Hume quote is from Hume (1779, 106–7).

Biology Since Darwin

This section shares sources with the previous section, "The Deep Roots of Evolutionary Ideas in Biology." The basics of inheritance are described in Bonduriansky and Day (2018). The study of eye coloration in flies is from Morgan (1910). The core mathematical models in population genetics are described in textbooks, such as Hedrick (2011). Publications calling into question some of the assumptions of neo-Darwinism, with accounts of epigenetics and the extended evolutionary synthesis, include Jablonka and Lamb (2014), Laland et al. (2015), and Bonduriansky and Day (2018). For horizontal gene transfer in bacteria, see Brito (2021). For a broader treatment of the ways in which microbes challenge neo-Darwinian evolutionary theory, see Doolittle and Brunet (2016) and Inkpen and Doolittle (2022). The heritability of animal behavior is covered by Dochtermann et al. (2019); Mesoudi (2011) describes cases of cultural inheritance in nonhuman (and human) animals.

Cultural Evolution Since Darwin

Specific contributions mentioned in the text are Butler (1863, "Darwin Among the Machines"), Darwin (1871, *Descent of Man*), Veblen (1898, "Why Is Economics Not an Evolutionary Science?"), and Baldwin (1909, *Darwin and the Humanities*). The background and history of the field of cultural evolution are described in Blute (2010), Hodgson and Knudsen (2010), Mesoudi (2011), and Tang (2020). Evolutionary economics is covered by Schotter (1981), Nelson and Winter (1982), and Beinhocker (2006). Turchin (2003) applies evolutionary-ecological concepts to explain history. Spencer (1864) coined the term *survival of the fittest*, and Spencer (1865) used the expression *law of evolution*. The notes for "The Darwinian Distraction" section in chapter 1 list references for social Darwinism. Schumpeter (1942) describes "creative destruction" in economics. For game theory, see Maynard Smith (1982). Dawkins (1976) used the terms "replicator" and "vehicle" for the analogs of genes and organisms, and he coined the term *Universal Darwinism* (1983). Lewens (2015) argues against the need for the replicator-interactor dualism. The two books applying mathematical models to cultural evolution are Cavalli-Sforza and Feldman (1981) and Boyd and Richerson (1985).

Chapter 3

Introductory Text

The evolution of the English language is the subject of Crystal (2010) and Bryson (1990); sources for more general evolutionary approaches to language were provided in chapters 1 and 2.

Evolutionary Systems

One of the best-known versions of the three-ingredient recipe for evolutionary systems was presented by Lewontin (1970), although other versions have been published as well, some many decades earlier (reviewed in Godfrey-Smith 2009). Sherwin (2024) presents a version with four processes (similar to those presented in this book), applied to biology and to artificial intelligence, under the unifying theme of the "evolution of information."

Populations, Entities, and Success

Lloyd (2022) describes slime mold biology. The evolution of single entities is explored for the case of firms by Nelson et al. (2018). The scenario of cell division followed by immediate death of the parent is used by Godfrey-Smith (2009), who also shows that many concepts in evolutionary science defy precise definitions.

Evolution at Multiple Levels

Okasha (2006) provides an overview of multilevel selection and evolution. Population genetic models based on Mendel's laws are covered in Hedrick (2011); the application of similar models to ecological communities is presented in Vellend (2016). Bhattacharyya et al. (1990) describe the molecular biology behind wrinkly peas, and Kreplak et al. (2019) report on the whole pea genome. Estimates of the quantity of nonfunctional DNA in eukaryotes are reported in Palazzo and Gregory (2014). The story of Lynn Margulis's hypothesis of the endosymbiotic origin of mitochondria and chloroplasts is told in Gray (2017). The diversity and phylogeny of angiosperms are described in Stevens (2017) and Zomlefer (1994), while the phylogeny of the pea family (in fig. 3.5) is taken from Shatskaya et al. (2019). Data on pea yields are from Ritchie et al. (2023); the number of countries reporting peas (and other crops) as a commodity is presented in Khoury et al. (2014). *Joy of Cooking* (1997) is by Rombauer et al. Campbell (1965) refers to variants under selection that are not in existence at the same moment of time as "variations across occasions." Carlson (2000) provides a study of Thomas Edison's sketches of telephones.

Chapter 4

Introductory Text

Google's origin story and the quotes from Larry Page are reported in Isaacson (2014). Other stories of happenstance in the evolution of culture and technology—including antibiotics and steam engines—come from several sources (Basalla 1988, Ridley 2020). The original paper on the CRISPR-Cas9 technology is Jinek et al. (2012), with the larger story told in Isaacson (2021). The basic biology of CRISPR is explained in Koonin (2019) and Ledford (2017). Several studies report reduced mutation rates in functionally important parts of the genome (Martincorena et al. 2012, Frigola et al. 2017, Monroe et al. 2022). The study on the genetics of fruit-fly ribosomes is Aldrich and Maggert (2015). Overviews of nonrandom generation of genetic variation are provided by Jablonka and Lamb (2014) and Bonduriansky and Day (2018). Nicholl (2023) gives an overview and introduction to genetic engineering; Mackelprang and Lemaux (2020) cover plants specifically.

Variants Are Generated and Inherited in Many Ways

Merchant (2017) describes the evolution of the iPhone. The idea of tinkering or combining in evolution has appeared in multiple publications, sometimes with finer distinctions as well

(Jacob 1977, Basalla 1988, Simonton 1999, Boden 2009, Arthur 2011, Ridley 2020, Perry et al. 2021). Accounts of hybridization are provided by Bullini (1994) for animals and by Rieseberg (1997) for plants. The classification of modes of inheritance is from Jablonka and Lamb (2014). Cultural evolution in animals is treated by Whiten (2017, 2019). Lycett et al. (2016) describe the influence of the availability of local materials on tool design.

The Evolutionary Soundboard, Part I: Variation Generation

Facts about the *Oxford English Dictionary* come from the website https://www.oed.com/ as of 2023. Martin (2000) gives an account of Japanese swords. Lynch et al. (2016) report estimates of DNA mutation rates; some information on mutation in humans is also from Freeman and Herron (2015). Information on human births comes from The World Counts website (https://www.theworldcounts.com/). Epigenetic mutation rates are presented in Klironomos et al. (2013) and Hofmeister et al. (2020). The speciation rate for fish is from Rabosky et al. (2013). For cultural artifacts, such as pottery or weapons, Eerkens and Lipo (2005) and Premo (2020) model the distribution of traits (such as size) expected when variation is generated by random copying errors under various constraints. The concept of fitness is discussed in detail in Rosenberg and Bouchard (2015). The distribution shown in figure 4.3 is a stylized version of data presented in Eyre-Walker and Keightley (2007).

The Evolutionary Soundboard, Part II: Inheritance

The estimate of the percentage of bacterial genes involved in horizontal gene transfer comes from Dagan et al. (2008).

Chapter 5

Introductory Text

The process of early plant domestication is laid out (with some necessary speculation) by Diamond (1997). The story of the Green Revolution is told by Mann (2018). Pinhasi et al. (2008, 2015) studied the evolution of human diet and jaw morphology, while Aoki (1986) modeled the evolution of farming practices and lactose (in)tolerance. Bradshaw (2017) tells the story of plant domestication up to the present. The account of ant agriculture is from Schultz (2022) and Purugganan (2022), with the latter providing a broader treatment of domestication by nonhuman animals. Johnson and Munshi-South (2017) discuss urbanization as an agent of natural selection; Pigeon et al. (2016) describe hunting as artificial selection.

Selection: Systematic Differential Success

Definitions and quotes are taken from Earnshaw-Whyte (2018). The biology of the t haplotype in mice is described in Silver (1993); the term "gene drive" is used for this situation by Alphey et al. (2020) and Taylor and Ingvarsson (2003). Chudek et al. (2012) present a study of prestige bias in children's learning. Tang (2020) describes the influence of power on transmission in evolutionary systems. The debate about group selection is covered in Okasha (2006) and

Leigh Jr. (2010); Dawkins (1976) argues that apparent examples of group selection can be understood in terms of "selfish genes." Sexual selection is treated in most evolution textbooks, such as Freeman and Herron (2015). For sexual selection in human cultural evolution, see De Block and Dewitte (2007). The evolution of multicellularity is one of the major transitions in evolution described by Maynard Smith and Szathmáry (1997). Pagel (2012) discusses the idea of cultural evolution via one society displacing another; he also uses the expression "cultural survival vehicles." The evolutionary advantage of within-society cooperation is also treated in Henrich (2016) and Muthukrishna (2023).

The Evolutionary Soundboard, Part III: Differential Success—Selection

The violin study is presented in Nia et al. (2015). A study of disruptive selection on bird beak size is reported in Hendry et al. (2008). The speculation about selection on the length of works of fiction comes from some blog posts (e.g., one on springcedars.com called "Why are novellas rare in publishing?"), other online discussions, and impressions from my own experience; I was unable to find statistics on the number of works that are published each year of different lengths. Macnair (1987) reports strongly contrasting plant communities on soils with or without heavy metal pollution. An example of divergent cultural evolution in city neighborhoods is provided by Schroeder et al. (2014), with a broader review in Sampson et al. (2002). The result that spatial environmental variation has greater potential to support diversity than temporal environmental variation is a conclusion from population genetic models (Hedrick 2011). Statistics on baby names come from the US Social Security Administration website (http://www.ssa .gov) and the Statistics Canada website (https://www.statcan.gc.ca/). Newberry and Plotkin (2022) report frequency-dependent selection on baby name choices. Similar kinds of novelty biases for fashion and other trends are discussed in Acerbi et al. (2012) and Acerbi and Bentley (2014). Piertney and Oliver (2006) provide an overview of the major histocompatibility complex, and Spurgin and Richardson (2010) discuss the evidence for different mechanisms maintaining high diversity. Vellend (2016) gives an overview of how evolutionary processes maintain diversity in ecological communities. Several studies address the feedbacks that maintain the ecotone between forests dominated by sugar maple and balsam fir (Carteron et al. 2020, Averill et al. 2022, Tourville et al. 2023). Positive frequency-dependent selection was demonstrated for music preferences by Salganik et al. (2006) and for word use by Pagel et al. (2019).

Selection Is Complex . . . but Simple

The idea that evolutionary theory succeeds given its focus on the consequences rather than the causes of selection comes from Sober (1992).

Chapter 6

Introductory Text

The story of Tasmania and the idea of the collective brain (the latter covered in the next section) are taken from Henrich (2004, 2016), with a few details on construction materials gleaned from the Australian Museum website (https://www.australian.museum). MacArthur and Wilson

(1967) report the number of ants on different Melanesian islands; their data was specifically for the subfamily of ants called Ponerinae. Studies of color blindness in the Pingelap Atoll are reported in Sheffield (2000) and Stone and De Wilde (2018).

Drift

The simulations shown in figure 6.2 were generated with code from Online Box 2 in Vellend (2016). For nonnative species, including their effects on islands, see Sax et al. (2005). The instruction for tying shoelaces is from a page titled "How to Tie a Shoe" at Instructables (https:// www.instructables.com). For drift as a creative force in evolution, see Wagner (2019).

Movement

Details of the Polar Inuit story can be found in Boyd et al. (2011). The Arctic fox studies are reported in Hasselgren et al. (2018) and Norén et al. (2016). Bell et al. (2019) review studies of genetic rescue. The influence of movement on diversity is treated in books on population genetics (Hedrick 2011) or ecology (Vellend 2016). Terms used to describe the homogenizing effects of movement include "biotic homogenization" (Olden and Rooney 2006), "cultural convergence" (Kaasa and Minkov 2020), "globalization" (Holton 2000), and "McDonaldization" (Ritzer 1996).

Completing the Evolutionary Soundboard

For the idea of evolution as an algorithm, see Dennett (1996) and Lewens and Buskell (2023).

Chapter 7

Introductory Text

In terms of "bizarre behavior" exhibited by ChatGPT, one notable case is reported by Roose (2023).

The Evolution of Artificial Intelligence

I learned the basics of neural networks from New Scientist (2017b) and from videos by Grant Sanderson, listed in the references under the YouTube channel name 3Blue1Brown (2017a, 2017b). Orgel's second rule for biology is described in Dunitz and Joyce (2013).

Artificial Life and Genetic Engineering

Approaches and techniques in plant breeding are presented in Kingsbury (2009) and Schlegel (2018); a broader treatment of genetic engineering (including the story of Dolly the sheep) is provided by Hodge (2009). The original paper on the first transgenic bacteria is Cohen et al. (1973), with the backstory provided by an article entitled "Herbert W. Boyer and Stanley N. Cohen" on the website of the Science History Institute Museum and Library (https:// sciencehistory.org). The statement about more than 90% of leading crops coming from

genetically modified seeds is from the US Food and Drug Administration website (https://www
.fda.gov). Studies of bacteria engineered to fight other bacteria include Neil et al. (2021) and
Matteau et al. (2021); I also consulted Professor Sébastien Rodrigue at Université de Sherbrooke on some of the details and the broader context of these studies.

Memetic Engineering

Differences and preferences between Coke and Pepsi are reported in Van Doorn and Miloyan
(2018); specific differences in contents can be seen by comparing nutritional labels on the two
products. The history of Coca-Cola comes from a section of the company website (https://www
.coca-colacompany.com) entitled "History of Coca-Cola Advertising Slogans," and from Hays
(2005) and Wu (2017). General information on marketing and advertising, including the idea
of innovation-revolt cycles, is from Wu (2010, 2017). Acerbi (2020) reviews evidence for the
effectiveness of celebrity endorsements, finding the evidence equivocal. Examples of power
dynamics in nonhuman animals are from Dugatkin (2022).

Chapter 8

Introductory Text

An account of the causes of the financial crisis in 2008 is provided by the United States Financial
Crisis Inquiry Commission (2011) and summarized in many articles, including "2008 Reces-
sion: What It Was and What Caused It" at https://www.investopedia.com. Squartini et al.
(2013) present the study of lending patterns among banks. A similar conclusion is reached by
Battiston et al. (2016), who say that "small errors on the knowledge of the network of contracts
can lead to large errors in the probability of systemic defaults."

Balls Roll Downhill

Positive frequency dependent selection has not been definitively demonstrated for the deciduous-
coniferous transition, but several studies are strongly suggestive (Brice et al. 2020, Carteron
et al. 2020, Vellend et al. 2021). Studies of critical transitions and critical slowing down are
reported in Scheffer (2009) and Scheffer et al. (2021).

Ecosystems

Studies of critical transitions in coral reefs include Hughes (1994), Mumby et al. (2007), and
Mumby (2009); similar studies for shallow lakes are presented in Scheffer et al. (1993)
and Scheffer (2009).

Cultural Norms

The QWERTY story is drawn from several sources (David 1985, Liebowitz and Margolis 1990, Kay
2013, Şama 2014), including a YouTube video showing the functioning of an early typebasket

(https://www.youtube.com/watch?v=bGX0D1cHxgY&t=9s). Şama (2014) describes the Turkish Fğğıod system. Stango (2004) discusses the idea of lock-in of standards. Young (2015) covers the potential for critical transitions in social norms, including dueling and foot-binding. Both foot-binding and genital mutilation are addressed by Mackie and LeJeune (2009). The studies of assigning names to human faces are reported in Centola and Baronchelli (2015) and Centola et al. (2018).

Major Transitions in Evolution over the Long Term

The idea of major evolutionary transitions was proposed in a book by Maynard Smith and Szathmáry, *The Major Transitions in Evolution* (1997). A more recent perspective is provided in Herron (2021). The quote from Wilson is in *This View of Life* (2019). Similar ideas about the close link between competition and cooperation are presented in Turchin (2007), Pagel (2012), Henrich (2016), and Muthukrishna (2023). The idea that religion might be as much a consequence as a cause of war comes from Pagel (2012) and Choi and Bowles (2007). Turchin (2003) presents a quantitative approach to analyzing history, and the quote "the very stability . . ." comes from Turchin (2007).

Complex Adaptive Systems

Holland (2014) provides an introduction to complex systems, including complex adaptive systems. Turchin (2010) predicted increased political instability in the 2010s.

Chapter 9

Introductory Text

The notes for chapter 6 provide references for concepts and terms related to globalization. Most information about McDonald's restaurants—their numbers and offerings in different countries—is from the website https://corporate.mcdonalds.com or from country-specific links provided in its "Where We Operate" section (March 22, 2023). Nederveen Pieterse (2009) describes the ambiance in a Moscow McDonald's. I am not sure of the origin of the term "glocalization," but it is mentioned frequently in the literature (e.g., Nederveen Pieterse 2009, Steger 2020). Calculations of diet composition use data from the Food and Agricultural Organization as reported in Khoury et al. (2014); for "starchy carbohydrate sources," I summed values for maize, oats, potatoes, rice, wheat, sweet potatoes, rye, other roots, and yams. Martin et al. (2019) reported a similar study. Facts about the history of food plants come from Vaughan and Geissler (1997). For the number of different cultivars of apples, Elzebroek (2008) reports that there are more than 7,500. For tomatoes, several internet sources say there are more than 10,000 varieties; I found no scientific literature to support this specific number, but data files from the Tomato Genetics Resource Center (https://tgrc.ucdavis.edu, accessed on December 1, 2023) contained 3,633 unique codes for items in their collection ("accessions").

What Is Diversity?

Magurran (2004), Page (2010), and Vellend (2016) provide overviews of how diversity is measured.

Causes of Diversity

Contributions to increasing global diversity from hybridization (in the broad sense) are described by Thomas (2015, 2017) for biology and Shah (2020) for culture.

Geographic Patterns of Diversity

Consistent relationships between biological diversity and variables such as latitude, temperature, or island size are reported in Huston (1994) and Rosenzweig (1995). Hillebrand (2004) reviews latitudinal diversity gradients. The idea that speciation might happen faster in the tropics due to the speed of biochemical reactions is from Allen et al. (2002). The result that speciation is actually faster in the temperate zone than the tropics is from Weir and Schluter (2007). Svenning et al. (2015) describe the possible long-term impacts of glaciation on diversity. Stronger negative frequency-dependent selection among tropical trees is reported by LaManna et al. (2017). Collard and Foley (2002) report a latitudinal gradient in cultural diversity; for language diversity specifically, see Mace and Pagel (1995) and Hua et al. (2019). Greater religious diversity in larger populations is reported in Warf and Vincent (2007). Evidence for multiple processes (including environmental heterogeneity) underlying correlations between system size and diversity comes from numerous studies (Mace and Pagel 1995, Laitin et al. 2012, Gavin et al. 2013, Stein et al. 2014).

Diversity Change in the Anthropocene

The profound impact of humans on ecosystems over thousands of years is described by Mann (2005). For an overview of the idea of the Anthropocene, see Ellis (2018). The estimate of one to two species going extinct every million years is from Ceballos et al. (2015), who give an upper estimate of two, which is already rounded up from 1.8, which is itself higher than most previous estimates. Anthropocene extinctions of languages are reported in Loh and Harmon (2014); the specific numbers come from Simons (2019). Wurm (1991) details some of the causes of language loss. The rough estimate of the number of native tree species in Canada (140) is from Hosie (1979). For tree species in Montréal, I used a public database on August 9, 2022, retaining only entries with Latin species and genus names, correcting spelling errors, and eliminating redundancies. As of September 2023, the original website was not active but appeared at https://quebio.ca/fr/arbresmtl. Pautasso (2007) reports high biological diversity at the scale of whole cities. Increases in biodiversity on islands and in US states during the Anthropocene are detailed in Sax and Gaines (2003), and the general idea that local-scale diversity doesn't necessarily decline under human influence is laid out in Vellend (2017). Data on linguistic diversity in New York City is from Perlin et al. (2021). For all of North America, Mithun (1999) estimates approximately three hundred native languages. The GeoGuessr game is at https://www.geoguessr.com. Homogenization of floras across continents is described by Winter et al. (2010) and La Sorte et al. (2007). The study showing increased diversity of ecosystem types under human influence is Martins et al. (2022). Kaasa and Minkov (2020) report a suite of analyses of the World Values Survey data. The case of the apple maggot fly is reported in Filchak et al. (2000). Vellend et al. (2007) review cases in which introduced populations of plants and animals diverge genetically from native populations. The Light Warlpiri language is described in O'Shannessy (2005). Lightfoot (2006) provides a broader treatment of the recent emergence of new languages.

Valuing Diversity

Perspectives on the relationship between science and values with respect to biodiversity can be found in Newman et al. (2017), Vellend (2017), and Cardou and Vellend (2023). Trends for the Index of Linguistic Diversity are reported in Loh and Harmon (2014). Trends in biodiversity intactness indices are summarized in Purvis et al. (2019).

The Consequences of Diversity: Bonus or Penalty?

The consequences of diversity are reviewed by Page (2010, 2017), from whom I borrowed the term "bonus" to describe positive diversity effects. Ecological mechanisms underlying diversity effects are described by Cardinale et al. (2006), with largely positive effects reported in Cardinale et al. (2012). The results in figure 9.5 were redrawn from a graph presented in Pastore et al. (2021); the photograph was taken by Elizabeth Kleynhans and is used with permission. Vellend (2017) stresses the idea that it is more often the composition than the diversity of ecological communities that people value. Putnam (2007) describes the possibility of diversity eroding trust among people. Schimmelpfennig et al. (2022) discuss the "paradox of diversity." The correlation of American economic growth with birthplace diversity is reported in Docquier et al. (2020); a similar study for Africa is Easterly and Levine (1997). The general consensus that human diversity more often has positive than negative effects comes from multiple studies (Kemeny 2017, Page 2017, Ozgen 2021). The study of the striped bass fisheries in Massachusetts is Aminpour et al. (2021). The perspective that diversity effects tend to be relatively small is from Vellend (2017) and Docquier et al. (2020).

Chapter 10

Introductory Text

Mixtures of French and English are described in Papen (2014). Lindsey and Brown (2021) review the relationship between color and language. The "iron rule of explanation" as key to defining science comes from Strevens (2020); according to this definition, the humanities would not be part of science. The divide between the natural and social sciences is laid out in Kagan (2009). Gordon (1991) analyzes the history of the social sciences. Figure 10.1 is a simplified and rotated version of a diagram presented in Bernal (1939). The sources mentioned in the paragraph about other authors who have recognized physical and evolutionary processes as the fundamental pair in science are Bell (2007), Mayr (2004), Wilson (2019), and Kauffman (2019).

Overcoming the Darwinian Distraction

The quote "Darwinian nested within Super-Lamarckian inheritance" is from Tang (2020), who uses the term "ideational evolution" as a synonym (or nearly so) for "cultural evolution." The two books viewing evolutionary theory through different values-based lenses are Ridley (2015) and Wilson (2019).

The Faculty of Evolutionary Sciences

Resistance to the teaching of evolutionary ideas is covered in Futuyma (1995), with a recent example from India described in Shashidhara and Joshi (2023).

Choosing to See Commonalities or Differences

The quote "nothing in biology makes sense except in light of evolution" is from Dobzhansky (1973).

Did the Second Science Come First?

The idea of evolution in a population of universes is presented by Smolin (1997) and described in a popular article by Greene (2012), from whom I borrowed the language of clumps of matter being "blown apart" or "collapsing," and McFadden (2021).

REFERENCES

3Blue1Brown. 2017a. "But What Is a Neural Network?," chap. 1.

3Blue1Brown. 2017b. "Gradient Descent, How Neural Networks Learn," chap. 2.

Acerbi, A. 2020. *Cultural Evolution in the Digital Age*. Oxford University Press.

Acerbi, A., and R. A. Bentley. 2014. "Biases in Cultural Transmission Shape the Turnover of Popular Traits." *Evolution and Human Behavior* 35: 228–36.

Acerbi, A., S. Ghirlanda, and M. Enquist. 2012. "The Logic of Fashion Cycles." *PLoS One* 7: e32541.

Aldrich, J. C., and K. A. Maggert. 2015. "Transgenerational Inheritance of Diet-Induced Genome Rearrangements in *Drosophila*." *PLoS Genetics* 11: e1005148.

Allen, A. P., J. H. Brown, and J. F. Gillooly. 2002. "Global Biodiversity, Biochemical Kinetics, and the Energetic-Equivalence Rule." *Science* 297: 1545–48.

Alphey, L. S., A. Crisanti, F. Randazzo, and O. S. Akbari. 2020. "Standardizing the Definition of Gene Drive." *Proceedings of the National Academy of Sciences USA* 117: 30864–67.

Alter, S. G. 2003. *Darwinism and the Linguistic Image*. Johns Hopkins University Press.

Aminpour, P., S. A. Gray, A. Singer et al. 2021. "The Diversity Bonus in Pooling Local Knowledge About Complex Problems." *Proceedings of the National Academy of Sciences USA* 118: e2016887118.

Aoki, K. 1986. "A Stochastic Model of Gene-Culture Coevolution Suggested by the 'Culture Historical Hypothesis' for the Evolution of Adult Lactose Absorption in Humans." *Proceedings of the National Academy of Sciences USA* 83: 2929–33.

Arthur, W. B. 2011. *The Nature of Technology: What It Is and How It Evolves*. Free Press.

Averill, C., C. Fortunel, D. S. Maynard et al. 2022. "Alternative Stable States of the Forest Mycobiome Are Maintained Through Positive Feedbacks." *Nature Ecology & Evolution* 6: 375–82.

Baldwin, J. M. 1909. *Darwin and the Humanities*. Review Pub. Co.

Basalla, G. 1988. *The Evolution of Technology*. Cambridge University Press.

Battiston, S., G. Caldarelli, R. M. May, T. Roukny, and J. E. Stiglitz. 2016. "The Price of Complexity in Financial Networks." *Proceedings of the National Academy of Sciences USA* 113: 10031–36.

Beinhocker, E. D. 2006. *The Origin of Wealth: Evolution, Complexity, and the Radical Remaking of Economics*. Harvard Business Review Press.

Bell, D. A., Z. L. Robinson, W. C. Funk et al. 2019. "The Exciting Potential and Remaining Uncertainties of Genetic Rescue." *Trends in Ecology & Evolution* 34: 1070–79.

Bell, G. 2007. *Selection: The Mechanism of Evolution.* Oxford University Press.

Bernal, J. D. 1939. *The Social Function of Science.* Routledge.

Bhattacharyya, M. K., A. M. Smith, T.H.N. Ellis, C. Hedley, and C. Martin. 1990. "The Wrinkled-Seed Character of Pea Described by Mendel Is Caused by a Transposon-like Insertion in a Gene Encoding Starch-Branching Enzyme." *Cell* 60: 115–22.

Blute, M. 2010. *Darwinian Sociocultural Evolution: Solutions to Dilemmas in Cultural and Social Theory.* Cambridge University Press.

Blute, M. 2020. "Gene-Culture and Potential Culture-Gene Coevolution: The Future of COVID-19." *This View of Life Magazine,* July 22, 2024. https://www.prosocial.world/posts /gene-culture-and-potential-culture-gene-coevolution-the-future-of-covid-19.

Boden, M. A. 2009. "Computer Models of Creativity." *AI Magazine* 30: 23–23.

Bonduriansky, R., and T. Day. 2018. *Extended Heredity: A New Understanding of Inheritance and Evolution.* Princeton University Press.

Boyd, R., and P. J. Richerson. 1985. *Culture and the Evolutionary Process.* University of Chicago Press.

Boyd, R., P. J. Richerson, and J. Henrich. 2011. "The Cultural Niche: Why Social Learning Is Essential for Human Adaptation." *Proceedings of the National Academy of Sciences USA* 108: 10918–25.

Bradshaw, J. E. 2017. "Plant Breeding: Past, Present and Future." *Euphytica* 213: 60.

Brice, M-H, S. Vissault, W. Vieira, D. Gravel, P. Legendre, and M-J. Fortin. 2020. "Moderate Disturbances Accelerate Forest Transition Dynamics Under Climate Change in the Temperate– Boreal Ecotone of Eastern North America." *Global Change Biology* 26: 4418–35.

Brito, I. L. 2021. "Examining Horizontal Gene Transfer in Microbial Communities." *Nature Reviews Microbiology* 19: 442–53.

Bryson, B. 1990. *The Mother Tongue: English and How It Got That Way.* HarperCollins.

Buchanan, M. 2023. "Science Strengthened Banks—but How Long Will Stability Last?" *Nature* 616: 642–42.

Bullini, L. 1994. "Origin and Evolution of Animal Hybrid Species." *Trends in Ecology & Evolution* 9: 422–26.

Butler, S. 1863. *Darwin Among the Machines.* The Press.

Campbell, D. T. 1965. "Variation and Selective Retention in Socio-cultural Evolution." In *Social Change in Developing Areas: A Reinterpretation of Evolutionary Theory,* edited by H. R. Barringer, G. I. Blankstein, and R. W. Mack, 19–48. Schenkman.

Campbell, L. 2020. *Historical Linguistics: An Introduction.* 4th ed. MIT Press.

Cardinale, B. J., J. E. Duffy, A. Gonzalez et al. 2012. "Biodiversity Loss and Its Impact on Humanity." *Nature* 486: 59–67.

Cardinale, B. J., D. S. Srivastava, J. Emmett Duffy et al. 2006. "Effects of Biodiversity on the Functioning of Trophic Groups and Ecosystems." *Nature* 443: 989–92.

Cardou, F., and M. Vellend. 2023. "Stealth Advocacy in Ecology and Conservation Biology." *Biological Conservation* 280: 109968.

Carlson, W. B. 2000. "Invention and Evolution: The Case of Edison's Sketches of the Telephone." In *Technological Innovation as an Evolutionary Process,* edited by J. M. Ziman, 137–58. Cambridge University Press.

Carteron, A., V. Parasquive, F. Blanchard, et al. 2020. "Soil Abiotic and Biotic Properties Constrain the Establishment of a Dominant Temperate Tree into Boreal Forests." *Journal of Ecology* 108: 931–44.

Cavalli-Sforza, L. L., and M. W. Feldman. 1981. *Cultural Transmission and Evolution: A Quantitative Approach.* Princeton University Press.

Ceballos, G., P. R. Ehrlich, A. D. Barnosky, A. García, R. M. Pringle, and R. M. Palmer. 2015. "Accelerated Modern Human–Induced Species Losses: Entering the Sixth Mass Extinction." *Science Advances* 1: e1400253.

Centola, D., and A. Baronchelli. 2015. "The Spontaneous Emergence of Conventions: An Experimental Study of Cultural Evolution." *Proceedings of the National Academy of Sciences USA* 112: 1989–94.

Centola, D., J. Becker, D. Brackbill, and A. Baronchelli. 2018. "Experimental Evidence for Tipping Points in Social Convention." *Science* 360: 1116–19.

Choi, J-K, and S. Bowles. 2007. "The Coevolution of Parochial Altruism and War." *Science* 318: 636–40.

Chudek, M., S. Heller, S. Birch, and J. Henrich. 2012. "Prestige-Biased Cultural Learning: Bystander's Differential Attention to Potential Models Influences Children's Learning." *Evolution and Human Behavior* 33: 46–56.

Cohen, S. N., A.C.Y. Chang, H. W. Boyer, and R. B. Helling. 1973. "Construction of Biologically Functional Bacterial Plasmids in Vitro." *Proceedings of the National Academy of Sciences USA* 70: 3240–44.

Collard, I. F., and R. A. Foley. 2002. "Latitudinal Patterns and Environmental Determinants of Recent Human Cultural Diversity: Do Humans Follow Biogeographical Rules?" *Evolutionary Ecology Research* 4: 371–83.

Cook, A. H. 1998. *Edmond Halley: Charting the Heavens and the Seas.* Clarendon Press.

Crystal, D. 2010. *Evolving English: One Language, Many Voices: An Illustrated History of the English Language.* British Library.

Cziko, G. 1997. *Without Miracles: Universal Selection Theory and the Second Darwinian Revolution.* MIT Press.

Dagan, T., Y. Artzy-Randrup, and W. Martin. 2008. "Modular Networks and Cumulative Impact of Lateral Transfer in Prokaryote Genome Evolution." *Proceedings of the National Academy of Sciences USA* 105: 10039–44.

Darwin, C. 1859. *On the Origin of Species by Means of Natural Selection, Or, The Preservation of Favoured Races in the Struggle for Life.* J. Murray.

Darwin, C. 1871. *The Descent of Man.* J. Murray.

David, P. A. 1985. "Clio and the Economics of QWERTY." *American Economic Review* 75: 332–37.

Dawkins, R. 1976. *The Selfish Gene.* Oxford University Press.

Dawkins, R. 1983. *Universal Darwinism. Evolution from Molecules to Men,* edited by D. S. Bendall, 403–25. Cambridge University Press.

De Block, A, and S. Dewitte. 2007. "Mating Games: Cultural Evolution and Sexual Selection." *Biology & Philosophy* 22: 475–91.

Dennett, D. C. 1996. *Darwin's Dangerous Idea: Evolution and the Meaning of Life.* Simon and Schuster.

Diamond, J. 1997. *Guns, Germs, and Steel: The Fates of Human Societies*. Norton.

Dobzhansky, T. 1973. "Nothing in Biology Makes Sense Except in the Light of Evolution." *American Biology Teacher* 35: 125–29.

Dochtermann, N. A., T. Schwab, M. Anderson Berdal, J. Dalos, and R. Royauté. 2019. "The Heritability of Behavior: A Meta-analysis." *Journal of Heredity* 110: 403–10.

Docquier, F., R. Turati, J. Valette, and C. Vasilakis. 2020. "Birthplace Diversity and Economic Growth: Evidence from the US States in the Post–World War II Period." *Journal of Economic Geography* 20: 321–54.

Doolittle, W. F., and T.D.P. Brunet. 2016. "What Is the Tree of Life?" *PLoS Genetics* 12: e1005912.

Dugatkin, L. A. 2022. *Power in the Wild: The Subtle and Not-So-Subtle Ways Animals Strive for Control over Others*. University of Chicago Press.

Dunitz, J. D., and G. F. Joyce. 2013. *Biographical Memoirs: Leslie E. Orgel 1927–2007*. National Academy of Sciences.

Earnshaw-Whyte, E. 2018. *Modelling Evolution: A New Dynamic Account*. Routledge.

Easterly, W., and R. Levine. 1997. "Africa's Growth Tragedy: Policies and Ethnic Divisions." *Quarterly Journal of Economics* 112: 1203–50.

Eerkens, J. W., and C. P. Lipo. 2005. "Cultural Transmission, Copying Errors, and the Generation of Variation in Material Culture and the Archaeological Record." *Journal of Anthropological Archaeology* 24: 316–34.

Ellis, E. C. 2018. *Anthropocene: A Very Short Introduction*. Oxford University Press.

Elzebroek, A.T.G. 2008. *Guide to Cultivated Plants*. CABI.

Eyre-Walker, A., and P. D. Keightley. 2007. "The Distribution of Fitness Effects of New Mutations." *Nature Reviews Genetics* 8: 610–18.

Filchak, K. E., J. B. Roethele, and J. L. Feder. 2000. "Natural Selection and Sympatric Divergence in the Apple Maggot *Rhagoletis pomonella*." *Nature* 407: 739–42.

Freeman, S., and J. C. Herron. 2015. *Evolutionary Analysis*. Pearson.

Frigola, J., R. Sabarinathan, L. Mularoni, F. Muiños, A. Gonzalez-Perez, and N. López-Bigas. 2017. "Reduced Mutation Rate in Exons Due to Differential Mismatch Repair." *Nature Genetics* 49: 1684–92.

Futuyma, D. J. 1995. *Science on Trial: The Case for Evolution*. Sinauer.

Futuyma, D. J., and M. Kirkpatrick. 2017. *Evolution*. Sinauer.

Gavin, M. C., C. A. Botero, C. Bowern et al. 2013. "Toward a Mechanistic Understanding of Linguistic Diversity." *BioScience* 63: 524–35.

Godfrey-Smith, P. 2009. *Darwinian Populations and Natural Selection*. Oxford University Press.

Godfrey-Smith, P. 2014. *Philosophy of Biology*. Princeton University Press.

Gordon, H. S. 1991. *The History and Philosophy of Social Science*. Routledge.

Gould, S. J. 1980. *The Panda's Thumb: More Reflections in Natural History*. Norton.

Gray, M. W. 2017. "Lynn Margulis and the Endosymbiont Hypothesis: 50 Years Later." *Molecular Biology of the Cell* 28: 1285–87.

Greene, B. 2012. "Welcome to the Multiverse." *Newsweek*. https://www.newsweek.com/brian-greene-welcome-multiverse-64887.

Gribbin, J., and M. Gribbin. 2022. *On The Origin of Evolution: Tracing "Darwin's Dangerous Idea" from Aristotle to DNA*. Rowman & Littlefield.

Grosberg, R. K., and R. R. Strathmann. 2007. "The Evolution of Multicellularity: A Minor Major Transition?" *Annual Review of Ecology, Evolution, and Systematics* 38: 621–54.

Haldane, A. G, and R. M. May. 2011. "Systemic Risk in Banking Ecosystems." *Nature* 469: 351–55.

Hasselgren, M., A. Angerbjörn, N. E. Eide et al. 2018. "Genetic Rescue in an Inbred Arctic Fox (*Vulpes lagopus*) Population." *Proceedings of the Royal Society B: Biological Sciences* 285: 20172814.

Hays, C. L. 2005. *The Real Thing: Truth and Power at the Coca-Cola Company*. Random House.

Hedrick, P. 2011. *Genetics of Populations*. Jones & Bartlett Learning.

Hendry, A. P., S. K. Huber, L. F. De León, A. Herrel, and J. Podos. 2008. "Disruptive Selection in a Bimodal Population of Darwin's Finches." *Proceedings of the Royal Society B: Biological Sciences* 276: 753–59.

Henrich, J. 2004. "Demography and Cultural Evolution: How Adaptive Cultural Processes Can Produce Maladaptive Losses: The Tasmanian Case." *American Antiquity* 69: 197–214.

Henrich, J. 2016. *The Secret of Our Success: How Culture Is Driving Human Evolution, Domesticating Our Species, and Making Us Smarter*. Princeton University Press.

Herron, M. D. 2021. "What Are the Major Transitions?" *Biology & Philosophy* 36: 2.

Hillebrand, H. 2004. "On the Generality of the Latitudinal Diversity Gradient." *American Naturalist* 163: 192–211.

Hodge, R. 2009. *Genetic Engineering: Manipulating the Mechanisms of Life*. Infobase Publishing.

Hodgson, G. M., and T. Knudsen. 2010. *Darwin's Conjecture: The Search for General Principles of Social and Economic Evolution*. University of Chicago Press.

Hofmeister, B. T., J. Denkena, M. Colomé-Tatché et al. 2020. "A Genome Assembly and the Somatic Genetic and Epigenetic Mutation Rate in a Wild Long-lived Perennial *Populus trichocarpa*." *Genome Biology* 21: 259.

Holland, J. H. 2014. *Complexity: A Very Short Introduction*. Oxford University Press.

Holton, R. 2000. "Globalization's Cultural Consequences." *Annals of the American Academy of Political and Social Science* 570: 140–52.

Hosie, R. C. 1979. *Native Trees of Canada*. 8th ed. Fitzhenry & Whiteside.

Hua, X., S. J. Greenhill, M. Cardillo, H. Schneemann, and L. Bromham. 2019. "The Ecological Drivers of Variation in Global Language Diversity." *Nature Communications* 10: 1–10.

Hughes, T. P. 1994. "Catastrophes, Phase Shifts, and Large-scale Degradation of a Caribbean Coral Reef." *Science* 265: 1547–51.

Hull, D. L. 1988. *Science as a Process: An Evolutionary Account of the Social and Conceptual Development of Science*. University of Chicago Press.

Hume, D. 1779. *Dialogues Concerning Natural Religion*. Penguin Books.

Huston, M. A. 1994. *Biological Diversity: The Coexistence of Species*. Cambridge University Press.

Huxley, J. 1942. *Evolution: The Modern Synthesis*. Allen & Unwin.

Inkpen, S. A., and W. F. Doolittle. 2022. *Can Microbial Communities Regenerate? Uniting Ecology and Evolutionary Biology*. University of Chicago Press.

Isaacson, W. 2014. *The Innovators: How a Group of Hackers, Geniuses, and Geeks Created the Digital Revolution*. Simon and Schuster.

Isaacson, W. 2021. *The Code Breaker: Jennifer Doudna, Gene Editing, and the Future of the Human Race*. Simon and Schuster.

Jablonka, E., and M. J. Lamb. 2014. *Evolution in Four Dimensions: Genetic, Epigenetic, Behavioral, and Symbolic Variation in the History of Life*. MIT Press.

Jacob, F. 1977. "Evolution and Tinkering." *Science* 196: 1161–66.

Jinek, M., K. Chylinski, I. Fonfara, M. Hauer, J. A. Doudna, and E. Charpentier. 2012. "A Programmable Dual-RNA–Guided DNA Endonuclease in Adaptive Bacterial Immunity." *Science* 337: 816–21.

Johnson, M.T.J., and J. Munshi-South. 2017. "Evolution of Life in Urban Environments." *Science* 358: eaam8327.

Jones, W. 1799. *The Works of Sir William Jones in Six Volumes*. Volume 1. G. G. and J. Robinson, Paternoster Row.

Kaasa, A., and M. Minkov. 2020. "Are the World's National Cultures Becoming More Similar?" *Journal of Cross-Cultural Psychology* 51: 531–50.

Kagan, J. 2009. *The Three Cultures: Natural Sciences, Social Sciences, and the Humanities in the 21st Century*. Cambridge University Press.

Kauffman, S. 2019. *A World Beyond Physics*. Oxford University Press.

Kay, N. M. 2013. "Rerun the Tape of History and QWERTY Always Wins." *Research Policy* 42: 1175–85.

Kemeny, T. 2017. "Immigrant Diversity and Economic Performance in Cities." *International Regional Science Review* 40: 164–208.

Khoury, C. K., A. D. Bjorkman, H. Dempewolf, et al. 2014. "Increasing Homogeneity in Global Food Supplies and the Implications for Food Security." *Proceedings of the National Academy of Sciences USA* 111: 4001–4006.

Kingsbury, N. 2009. *Hybrid: The History and Science of Plant Breeding*. University of Chicago Press.

Klironomos, F. D., J. Berg, and S. Collins. 2013. "How Epigenetic Mutations Can Affect Genetic Evolution: Model and Mechanism." *BioEssays* 35: 571–78.

Koonin, E. V. 2019. "CRISPR: A New Principle of Genome Engineering Linked to Conceptual Shifts in Evolutionary Biology." *Biology & Philosophy* 34: 9.

Kreplak, J., M-A Madoui, P. Cápal et al. 2019. "A Reference Genome for Pea Provides Insight into Legume Genome Evolution." *Nature Genetics* 51: 1411–22.

La Sorte, F. A., M. L. McKinney, and P. Pyšek. 2007. "Compositional Similarity Among Urban Floras Within and Across Continents: Biogeographical Consequences of Human-Mediated Biotic Interchange." *Global Change Biology* 13: 913–21.

Laitin, D. D., J. Moortgat, and A. L. Robinson. 2012. "Geographic Axes and the Persistence of Cultural Diversity." *Proceedings of the National Academy of Sciences USA* 109: 10263–68.

Laland, K. N., T. Uller, M. W. Feldman et al. 2015. "The Extended Evolutionary Synthesis: Its Structure, Assumptions and Predictions." *Proceedings of the Royal Society B: Biological Sciences* 282: 20151019.

LaManna, J. A., S. A. Mangan, A. Alonso et al. 2017. "Plant Diversity Increases with the Strength of Negative Density Dependence at the Global Scale." *Science* 356: 1389–92.

Lamarck, J-B P. A. 1809. "Philosophie Zoologique." *Dentu et L'Auteur*.

Lavrov, A. I., and I. A. Kosevich. 2014. "Sponge Cell Reaggregation: Mechanisms and Dynamics of the Process." *Russian Journal of Developmental Biology* 45: 205–23.

Ledford, H. 2017. "Five Big Mysteries About CRISPR's Origins." *Nature* 541: 280–82.

Leigh Jr., E. G. 2010. "The Group Selection Controversy." *Journal of Evolutionary Biology* 23: 6–19.

Lewens, T. 2015. *Cultural Evolution: Conceptual Challenges.* Oxford University Press.

Lewens, T., and A. Buskell. 2023. "Cultural Evolution." In *The Stanford Encyclopedia of Philosophy*, edited by E. N. Zalta and U. Nodelman, Summer 2023. Metaphysics Research Lab, Stanford University.

Lewontin, R. C. 1970. "The Units of Selection." *Annual Review of Ecology and Systematics* 1: 1–18.

Liebowitz, S. J., and S. E. Margolis. 1990. "The Fable of the Keys." *Journal of Law and Economics* 33: 1–25.

Lightfoot, D. 2006. *How New Languages Emerge.* Cambridge University Press.

Lindsey, D. T., and A. M. Brown. 2021. "Lexical Color Categories." *Annual Review of Vision Science* 7: 605–31.

Lloyd, S. 2022. *Where the Slime Mould Creeps.* Tympanocryptis Press.

Loh, J., and D. Harmon. 2014. *Biocultural Diversity: Threatened Species, Endangered Languages.* WWF Netherlands.

Lycett, S. J., K. Schillinger, M. I. Eren, N. von Cramon-Taubadel, and A. Mesoudi. 2016. "Factors Affecting Acheulean Handaxe Variation: Experimental Insights, Microevolutionary Processes, and Macroevolutionary Outcomes." *Quaternary International* 411: 386–401.

Lynch, M., M. S. Ackerman, J-F Gout et al. 2016. "Genetic Drift, Selection and the Evolution of the Mutation Rate." *Nature Reviews Genetics* 17: 704–14.

MacArthur, R. H., and E. O. Wilson. 1967. *The Theory of Island Biogeography.* Princeton University Press.

Mace, R., and M. Pagel. 1995. "A Latitudinal Gradient in the Density of Human Languages in North America." *Proceedings of the Royal Society B: Biological Sciences* 261: 117–21.

Mackelprang, R., and P. G. Lemaux. 2020. "Genetic Engineering and Editing of Plants: An Analysis of New and Persisting Questions." *Annual Review of Plant Biology* 71: 659–87.

Mackie, G., and J. LeJeune. 2009. *Social Dynamics of Abandonment of Harmful Practices: A New Look at the Theory.* UNICEF Innocenti Research Centre.

Macnair, M. R. 1987. "Heavy Metal Tolerance in Plants: A Model Evolutionary System." *Trends in Ecology & Evolution* 2: 354–59.

Magurran, A. E. 2004. *Measuring Biological Diversity.* Wiley.

Mann, C. C. 2005. *1491: New Revelations of the Americas Before Columbus.* Knopf Doubleday.

Mann, C. C. 2018. *The Wizard and the Prophet: Two Dueling Scientists and Their Dueling Visions to Shape Tomorrow's World.* Knopf Doubleday.

Martin, A. R., M. W. Cadotte, M. E. Isaac, R. Milla, D. Vile, and C. Violle. 2019. "Regional and Global Shifts in Crop Diversity Through the Anthropocene." *PLoS One* 14: e0209788.

Martin, G. 2000. "Stasis in Complex Artefacts." In *Technological Innovation as an Evolutionary Process*, edited by J. M. Ziman, 90–100. Cambridge University Press.

Martincorena, I., A.S.N. Seshasayee, and N. M. Luscombe. 2012. "Evidence of Non-random Mutation Rates Suggests an Evolutionary Risk Management Strategy." *Nature* 485: 95–98.

Martins, I. S., M. Dornelas, M. Vellend, and C. D. Thomas. 2022. "A Millennium of Increasing Diversity of Ecosystems Until the Mid-20th Century." *Global Change Biology* 28: 5945–55.

Matteau, D., and S. Rodrigue. 2021. "An Engineered *Mycoplasma pneumoniae* to Fight *Staphylococcus aureus*." *Molecular Systems Biology* 17: e10574.

Matthew, P. 1831. *On Naval Timber and Arboriculture: With Critical Notes on Authors Who Have Recently Treated the Subject of Planting*. A. Black.

Mayer, W. V. 1987. "Wallace and Darwin." *American Biology Teacher* 49: 406–10.

Maynard Smith, J. 1982. *Evolution and the Theory of Games*. Cambridge University Press.

Maynard Smith, J., and E. Szathmáry. 1997. *The Major Transitions in Evolution*. Oxford University Press.

Mayr, E. 1982. *The Growth of Biological Thought: Diversity, Evolution, and Inheritance*. Harvard University Press.

Mayr, E. 2004. *What Makes Biology Unique? Considerations on the Autonomy of a Scientific Discipline*. Cambridge University Press.

McFadden, J. 2021. "Why Simplicity Works: Does the Existence of a Multiverse Hold the Key for Why Nature's Laws Seem So Simple?" *Aeon*. https://aeon.co/essays/why-is-simplicity-so-unreasonably-effective-at-scientific-explanation.

McShea, D. W., and R. N. Brandon. 2010. *Biology's First Law: The Tendency for Diversity and Complexity to Increase in Evolutionary Systems*. University of Chicago Press.

Merchant, B. 2017. *The One Device: The Secret History of the iPhone*. Little, Brown.

Mesoudi, A. 2011. *Cultural Evolution: How Darwinian Theory Can Explain Human Culture and Synthesize the Social Sciences*. University of Chicago Press.

Metcalfe, J. S. 1998. *Evolutionary Economics and Creative Destruction*. Routledge.

Mithun, M. 1999. *The Languages of Native North America*. Cambridge University Press.

Monroe, J. G., T. Srikant, P. Carbonell-Bejerano et al. 2022. "Mutation Bias Reflects Natural Selection in *Arabidopsis thaliana*." *Nature* 602: 101–5.

Morgan, T. H. 1910. "Sex Limited Inheritance in *Drosophila*." *Science* 32: 120–22.

Mumby, P. J. 2009. "Phase Shifts and the Stability of Macroalgal Communities on Caribbean Coral Reefs." *Coral Reefs* 28: 761–73.

Mumby, P. J., A. Hastings, and H. J. Edwards. 2007. "Thresholds and the Resilience of Caribbean Coral Reefs." *Nature* 450: 98–101.

Muthukrishna, M. 2023. *A Theory of Everyone: The New Science of Who We Are, How We Got Here, and Where We're Going*. MIT Press.

Nederveen Pieterse, J. 2009. *Globalization and Culture: Global Mélange*. Rowman & Littlefield.

Neil, K., N. Allard, P. Roy et al. 2021. "High-Efficiency Delivery of CRISPR-Cas9 by Engineered Probiotics Enables Precise Microbiome Editing." *Molecular Systems Biology* 17: e10335.

Nelson, R. R., G. Dosi, C. E. Helfat et al. 2018. *Modern Evolutionary Economics: An Overview*. Cambridge University Press.

Nelson, R. R., and S. G. Winter. 1982. *An Evolutionary Theory of Economic Change*. Harvard University Press.

New Scientist. 2017a. *How Evolution Explains Everything About Life: From Darwin's Brilliant Idea to Today's Epic Theory*. John Murray.

New Scientist. 2017b. *Machines That Think: Everything You Need to Know About the Coming Age of Artificial Intelligence*. John Murray.

Newberry, M. G., and J. B. Plotkin. 2022. "Measuring Frequency-Dependent Selection in Culture." *Nature Human Behaviour* 6: 1048–55.

Newman, J. A., G. Varner, and S. Linquist. 2017. *Defending Biodiversity: Environmental Science and Ethics*. Cambridge University Press.

Nia, H. T., A. D. Jain, Y. Liu, M-R Alam, R. Barnas, and N. C. Makris. 2015. "The Evolution of Air Resonance Power Efficiency in the Violin and Its Ancestors." *Proceedings of the Royal Society A: Mathematical, Physical and Engineering Sciences* 471: 20140905.

Nicholl, D.S.T. 2023. *An Introduction to Genetic Engineering*. 4th ed. Cambridge University Press.

Norén, K., E. Godoy, L. Dalén, T. Meijer, and A. Angerbjörn. 2016. "Inbreeding Depression in a Critically Endangered Carnivore." *Molecular Ecology* 25: 3309–18.

Odling-Smee, F. J., K. N. Lala, and M. W. Feldman. 2003. *Niche Construction: The Neglected Process in Evolution*. Princeton University Press.

Okasha, S. 2006. *Evolution and the Levels of Selection*. Clarendon Press.

Olden, J. D., and T. P. Rooney. 2006. "On Defining and Quantifying Biotic Homogenization." *Global Ecology and Biogeography* 15: 113–20.

O'Shannessy, C. 2005. "Light Warlpiri: A New Language." *Australian Journal of Linguistics* 25: 31–57.

Ozgen, C. 2021. "The Economics of Diversity: Innovation, Productivity and the Labour Market." *Journal of Economic Surveys* 35: 1168–1216.

Page, S. E. 2010. *Diversity and Complexity*. Princeton University Press.

Page, S. E. 2017. *The Diversity Bonus: How Great Teams Pay Off in the Knowledge Economy*. Princeton University Press.

Pagel, M. 2012. *Wired for Culture: Origins of the Human Social Mind*. Norton.

Pagel, M., M. Beaumont, A. Meade, A. Verkerk, and A. Calude. 2019. "Dominant Words Rise to the Top by Positive Frequency-Dependent Selection." *Proceedings of the National Academy of Sciences USA* 116: 7397–7402.

Palazzo, A. F., and T. R. Gregory. 2014. "The Case for Junk DNA." *PLoS Genetics* 10: e1004351.

Paley, W. 1802. *Natural Theology: Or, Evidences of the Existence and Attributes of the Deity, Collected from the Appearances of Nature*. Gregg International.

Papen, R. A. 2014. "Hybrid Languages in Canada Involving French: The Case of Michif and Chiac." *Journal of Language Contact* 7: 154–83.

Pastore, M. A, S. E. Hobbie, and P. B. Reich. 2021. "Sensitivity of Grassland Carbon Pools to Plant Diversity, Elevated CO_2, and Soil Nitrogen Addition over 19 Years." *Proceedings of the National Academy of Sciences USA* 118: e2016965118.

Paul, D. B. 2009. "Darwin, Social Darwinism and Eugenics." In *The Cambridge Companion to Darwin*, edited by J. Hodge and G. Redick, 219–45. 2nd ed. Cambridge University Press.

Pautasso, M. 2007. "Scale Dependence of the Correlation Between Human Population Presence and Vertebrate and Plant Species Richness." *Ecology Letters* 10: 16–24.

Perlin, R., D. Kaufman, M. Turin, M. Daurio, S. Craig, and J. Lampel. 2021. "Mapping Urban Linguistic Diversity in New York City: Motives, Methods, Tools, and Outcomes." *Language Documentation & Conservation* 15: 458.

Perry, S., A. Carter, M. Smolla et al. 2021. "Not by Transmission Alone: The Role of Invention in Cultural Evolution." *Philosophical Transactions of the Royal Society B: Biological Sciences* 376: 20200049.

Piantadosi, S. T. 2014. "Zipf's Word Frequency Law in Natural Language: A Critical Review and Future Directions." *Psychonomic Bulletin & Review* 21: 1112–30.

Piertney, S. B., and M. K. Oliver. 2006. "The Evolutionary Ecology of the Major Histocompatibility Complex." *Heredity* 96: 7–21.

Pigeon, G., M. Festa-Bianchet, D. W. Coltman, and F. Pelletier. 2016. "Intense Selective Hunting Leads to Artificial Evolution in Horn Size." *Evolutionary Applications* 9: 521–30.

Pinhasi, R., V. Eshed, and N. von Cramon-Taubadel. 2015. "Incongruity Between Affinity Patterns Based on Mandibular and Lower Dental Dimensions Following the Transition to Agriculture in the Near East, Anatolia and Europe." *PLoS One* 10: e0117301.

Pinhasi, R., V. Eshed, and P. Shaw. 2008. "Evolutionary Changes in the Masticatory Complex Following the Transition to Farming in the Southern Levant." *American Journal of Physical Anthropology* 135: 136–48.

Popper, K. R. 1959. *The Logic of Scientific Discovery*. Basic Books.

Premo, L. S. 2020. "Reconciling and Reassessing the Accumulated Copying Error Model's Population-Level Predictions for Continuous Cultural Traits." *American Anthropologist* 122: 771–83.

Purugganan, M. D. 2022. "What Is Domestication?" *Trends in Ecology & Evolution* 37: 663–71.

Purvis, A., Z. Molnár, D. Obura et al. 2019. "Chapter 2.2 Status and Trends—Nature." In *Global Assessment Report of the Intergovernmental Science-Policy Platform on Biodiversity and Ecosystem Services*, edited by E. S. Brondízio, J. Settele, S. Díaz, and H. T. Ngo. IPBES Secretariat.

Putnam, R. D. 2007. "E Pluribus Unum: Diversity and Community in the Twenty-first Century." 2006 Johan Skytte Prize Lecture. *Scandinavian Political Studies* 30: 137–74.

Rabosky, D. L., F. Santini, J. Eastman et al. 2013. "Rates of Speciation and Morphological Evolution Are Correlated Across the Largest Vertebrate Radiation." *Nature Communications* 4: 1958.

Ridley, M. 2015. *The Evolution of Everything: How New Ideas Emerge*. HarperCollins.

Ridley, M. 2020. *How Innovation Works: And Why It Flourishes in Freedom*. HarperCollins.

Rieseberg, L. H. 1997. "Hybrid Origins of Plant Species." *Annual Review of Ecology and Systematics* 28: 359–89.

Ritchie, H., P. Rosado, and M. Roser. 2023. "Agricultural Production. Data Adapted from Food and Agriculture Organization of the United Nations." *Our World in Data*. https://ourworldindata.org/grapher/pea-yields.

Ritzer, G. 1996. "The McDonaldization Thesis: Is Expansion Inevitable?" *International Sociology* 11: 291–308.

Rombauer, I. von S., M. R. Becker, E. Becker, and M. Guarnaschelli. 1997. *Joy of Cooking*. Scribner.

Roose, K. 2023. "A Conversation with Bing's Chatbot Left Me Deeply Unsettled." *New York Times*. https://www.nytimes.com/2023/02/16/technology/bing-chatbot-microsoft-chatgpt.html.

Rosenberg, A., and F. Bouchard. 2015. "Fitness." In *Stanford Encyclopedia of Philosophy*, edited by E. N. Zalta and U. Nodelman. Metaphysics Research Lab, Stanford University.

Rosenzweig, M. L. 1995. *Species Diversity in Space and Time*. Cambridge University Press.

Salganik, M. J., P. S. Dodds, and D. J. Watts. 2006. "Experimental Study of Inequality and Unpredictability in an Artificial Cultural Market." *Science* 311: 854–56.

Şama, I. Y. 2014. "Economics of Qwerty and Fgğiod." *Journal of Economics Bibliography* 1: 17–25.

Sampson, R. J., J. D. Morenoff, and T. Gannon-Rowley. 2002. "Assessing 'Neighborhood Effects': Social Processes and New Directions in Research." *Annual Review of Sociology* 28: 443–78.

Sax, D. F., and S. D. Gaines. 2003. "Species Diversity: From Global Decreases to Local Increases." *Trends in Ecology & Evolution* 18: 561–66.

Sax, D. F., J. J. Stachowicz, and S. D. Gaines. 2005. *Species Invasions: Insights Into Ecology, Evolution, and Biogeography*. Freeman.

Scheffer, M. 2009. *Critical Transitions in Nature and Society*. Princeton University Press.

Scheffer, M., S. H. Hosper, M-L Meijer, B. Moss, and E. Jeppesen. 1993. "Alternative Equilibria in Shallow Lakes." *Trends in Ecology & Evolution* 8: 275–79.

Scheffer, M., E. H. van Nes, D. Bird, R. K. Bocinsky, and T. A. Kohler. 2021. "Loss of Resilience Preceded Transformations of pre-Hispanic Pueblo Societies." *Proceedings of the National Academy of Sciences USA* 118: e2024397118.

Schimmelpfennig, R., L. Razek, E. Schnell, and M. Muthukrishna. 2022. "Paradox of Diversity in the Collective Brain." *Philosophical Transactions of the Royal Society B: Biological Sciences* 377: 20200316.

Schlegel, R.H.J. 2018. *History of Plant Breeding*. CRC Press.

Schotter, A. 1981. *The Economic Theory of Social Institutions*. Cambridge University Press.

Schroeder, K. B., G. V. Pepper, and D. Nettle. 2014. "Local Norms of Cheating and the Cultural Evolution of Crime and Punishment: A Study of Two Urban Neighborhoods." *PeerJ* 2: e450.

Schultz, T. R. 2022. "The Convergent Evolution of Agriculture in Humans and Fungus-Farming Ants." In *The Convergent Evolution of Agriculture in Humans and Insects*, edited by T. R. Schultz, R. Gawne, and P. N. Peregrine, 281–313. MIT Press.

Schumpeter, J. A. 1942. *Capitalism, Socialism, and Democracy*. Harper and Brothers.

Shah, S. 2020. *The Next Great Migration: The Beauty and Terror of Life on the Move*. Bloomsbury Publishing USA.

Shashidhara, L. S., and A. Joshi. 2023. "Not Teaching Evolution Is an Injustice." *Science* 380: 1303.

Shatskaya, N., V. Bogdanova, O. Kosterin et al. 2019. "The Plastid and Mitochondrial Genomes of *Vavilovia formosa* (Stev.) Fed. and the Phylogeny of Related Legume Genera." *Vavilov Journal of Genetics and Breeding* 23: 972–80.

Sheffield, V. C. 2000. "The Vision of Typhoon Lengkieki." *Nature Medicine* 6: 746–47.

Sherwin, W. B. 2024. "Pan-evo: The Evolution of Information and Biology's Part in This." *Biology* 13: 507.

Silver, L. M. 1993. "The Peculiar Journey of a Selfish Chromosome: Mouse t Haplotypes and Meiotic Drive." *Trends in Genetics* 9: 250–54.

Simons, G. F. 2019. "Two Centuries of Spreading Language Loss." *Proceedings of the Linguistic Society of America* 4: 27: 1–12.

Simonton, D. K. 1999. "Creativity as Blind Variation and Selective Retention: Is the Creative Process Darwinian?" *Psychological Inquiry* 10: 309–28.

Smith, A. 1776. *An Inquiry Into the Nature and Causes of the Wealth of Nations.* W. Strahan and T. Cadell.

Smolin, L. 1997. *The Life of the Cosmos.* Oxford University Press.

Sober, E. 1992. "Models of Cultural Evolution." In *Trees of Life: Essays in Philosophy of Biology,* edited by P. Griffiths, 17–39. Springer Netherlands.

Spencer, H. 1864. *The Principles of Biology.* Williams.

Spencer, H. 1865. *First Principles of a New System of Philosophy.* Appleton.

Spurgin, L. G., and D. S. Richardson. 2010. "How Pathogens Drive Genetic Diversity: MHC, Mechanisms and Misunderstandings." *Proceedings of the Royal Society B: Biological Sciences* 277: 979–88.

Squartini, T., I. van Lelyveld, and D. Garlaschelli. 2013. "Early-Warning Signals of Topological Collapse in Interbank Networks." *Scientific Reports* 3: 3357.

Stango, V. 2004. "The Economics of Standards Wars." *Review of Network Economics* 3: 1–19.

Steger, M. B. 2020. *Globalization: A Very Short Introduction.* Oxford University Press.

Stein, A., K. Gerstner, and H. Kreft. 2014. "Environmental Heterogeneity as a Universal Driver of Species Richness Across Taxa, Biomes and Spatial Scales." *Ecology Letters* 17: 866–80.

Stevens, P. F. 2017. Angiosperm Phylogeny website. https://www.mobot.org/mobot/research/apweb/.

Stone, D., and S. De Wilde. 2018. "On Island of the Colorblind, Paradise Has a Different Hue." *National Geographic.* https://www.nationalgeographic.com/photography/article/pingelap-island-colorblindness-micronesia.

Stott, R. 2013. *Darwin's Ghosts: The Secret History of Evolution.* Random House.

Strevens, M. 2020. *The Knowledge Machine: How Irrationality Created Modern Science.* Liveright.

Svenning, J-C, W. L. Eiserhardt, S. Normand, A. Ordonez, and B. Sandel. 2015. "The Influence of Paleoclimate on Present-Day Patterns in Biodiversity and Ecosystems." *Annual Review of Ecology, Evolution, and Systematics* 46: 551–72.

Tang, S. 2020. *On Social Evolution: Phenomenon and Paradigm.* Routledge.

Taylor, D. R., and P. K. Ingvarsson. 2003. "Common Features of Segregation Distortion in Plants and Animals." *Genetica* 117: 27–35.

Thomas, C. D. 2015. "Rapid Acceleration of Plant Speciation During the Anthropocene." *Trends in Ecology & Evolution* 30: 448–55.

Thomas, C. D. 2017. *Inheritors of the Earth: How Nature Is Thriving in an Age of Extinction.* PublicAffairs.

Tourville, J. C., M. R. Zarfos, G. B. Lawrence, T. C. McDonnell, T. J. Sullivan, and M. Dovčiak. 2023. "Soil Biotic and Abiotic Thresholds in Sugar Maple and American Beech Seedling Establishment in Forests of the Northeastern United States." *Plant and Soil* 491: 387–400.

Turchin, P. 2003. *Historical Dynamics.* Princeton University Press.

Turchin, P. 2007. *War and Peace and War: The Rise and Fall of Empires.* Penguin Group.

Turchin, P. 2010. "Political Instability May Be a Contributor in the Coming Decade." *Nature* 463: 608.

United States Financial Crisis Inquiry Commission. 2011. "The Financial Crisis Inquiry Report: Final Report of the National Commission on the Causes of the Financial and Economic Crisis in the United States." Retrieved from the Library of Congress.

USDA. National Agricultural Statistics Service. 2023. "Crop Production Historical Track Records." https://downloads.usda.library.cornell.edu/usda-esmis/files/c534fn92g/vh53z8522/g732fp51d/croptr23.txt.

Van Doorn, G., and B. Miloyan. 2018. "The Pepsi Paradox: A Review." *Food Quality and Preference* 65: 194–97.

Van Wyhe, J. 2005. "The Descent of Words: Evolutionary Thinking 1780–1880." *Endeavour* 29: 94–100.

Van Wyhe, J. 2007. "Mind the Gap: Did Darwin Avoid Publishing His Theory for Many Years?" *Notes and Records of the Royal Society* 61: 177–205.

Vaughan, J. G., and C. A. Geissler. 1997. *The New Oxford Book of Food Plants.* Oxford University Press.

Veblen, T. 1898. "Why Is Economics Not an Evolutionary Science?" *Quarterly Journal of Economics* 12: 373–97.

Vellend, M. 2016. *The Theory of Ecological Communities.* Princeton University Press.

Vellend, M. 2017. "The Biodiversity Conservation Paradox." *American Scientist* 105: 94–101.

Vellend, M., M. Béhé, A. Carteron et al. 2021. "Plant Responses to Climate Change and an Elevational Gradient in Mont Mégantic National Park, Québec, Canada." *Northeastern Naturalist* 28: 4–28.

Vellend, M., L. J. Harmon, J. L. Lockwood et al. 2007. "Effects of Exotic Species on Evolutionary Diversification." *Trends in Ecology & Evolution* 22: 481–88.

Wagner, A. 2019. *Life Finds a Way: What Evolution Teaches Us About Creativity.* Basic Books.

Warf, B., and P. Vincent. 2007. "Religious Diversity Across the Globe: A Geographic Exploration." *Social and Cultural Geography* 8: 597–613.

Weir, J. T., and D. Schluter. 2007. "The Latitudinal Gradient in Recent Speciation and Extinction Rates of Birds and Mammals." *Science* 315: 1574–76.

Whewell, W. 1840. *Philosophy of the Inductive Sciences.* Vol. 2. John W. Parker.

Whiten, A. 2017. "A Second Inheritance System: The Extension of Biology Through Culture." *Interface Focus* 7: 20160142.

Whiten, A. 2019. "Cultural Evolution in Animals." *Annual Review of Ecology, Evolution, and Systematics* 50: 27–48.

Wilson, D. S. 2019. *This View of Life: Completing the Darwinian Revolution.* Knopf Doubleday.

Winter, M., I. Kühn, F.A.L. Sorte, O. Schweiger, W. Nentwig, and S. Klotz. 2010. "The Role of Non-native Plants and Vertebrates in Defining Patterns of Compositional Dissimilarity Within and Across Continents." *Global Ecology and Biogeography* 19: 332–42.

Witze, A. 2024. "It's Final: The Anthropocene Is Not an Epoch, Despite Protest over Vote." *Nature.* https://www.nature.com/articles/d41586-024-00868-1.

Wray, N, and P. Visscher. 2008. "Estimating Trait Heritability." *Nature Education* 1: 29.

Wu, C., W. Yin, Y. Jiang, and Xu HE. 2022. "Structure Genomics of SARS-CoV-2 and Its Omicron Variant: Drug Design Templates for COVID-19." *Acta Pharmacologica Sinica* 43: 3021–33.

Wu, T. 2010. *The Master Switch: The Rise and Fall of Information Empires*. Knopf Doubleday.

Wu, T. 2017. *The Attention Merchants: The Epic Scramble to Get Inside Our Heads*. Knopf Doubleday.

Wurm, S. A. 1991. "Language Death and Disappearance: Causes and Circumstances." *Diogenes* 39: 1–18.

Young, H. P. 2015. "The Evolution of Social Norms." *Annual Review of Economics* 7: 359–87.

Zipf, G. K. 1932. *Selected Studies of the Principles of Relative Frequency in Language*. Harvard University Press.

Zomlefer, W. B. 1994. *Guide to Flowering Plant Families*. University of North Carolina Press.

INDEX

Page numbers in *italics* refer to figures.